In Silico Dreams

In Silico Dreams

How Artificial Intelligence and Biotechnology Will Create the Medicines of the Future

Brian Hilbush

WILEY

About the Author

Brian Hilbush is a biotechnology and genomics expert with three decades of executive-level business and scientific management experience in the life sciences. Brian has worked at Fortune 500 companies, consulted for the pharmaceutical and semiconductor sectors, and commercialized technology platforms in genomics. Brian co-founded ModGene, where he built a genetics-based drug target discovery platform with a focus on neurodegenerative diseases. His technical expertise resides at the intersection of genomics, artificial intelligence, and bioinformatics. Brian has led technical teams developing software and analytics pipelines for human genomics studies, neuroanatomy, and gene expression. He has co-authored numerous scientific publications and was co-inventor on patents for novel treatments of Alzheimer's disease. Brian received his BS in cell and molecular biology from the University of Washington and his PhD in neuroscience from Stony Brook University in 1991. For more information on this book, please visit insilicodreams.com.

About the Technical Editor

Ted Coombs is a polymath futurist, technologist, and author of 24 technology books. His career in tech started in the 1970s as a laser engineer and roboticist. He began working in AI in 1983 for the Cleveland Clinic. He is one of the founders of modern computer forensics and today is also known for his knowledge of AIOps and DataOps. Ted is also an accomplished musician and fine artist.

Acknowledgments

This book was written entirely during the period spanning the COVID-19 pandemic. Prior to the outbreak, I discussed ideas for a book with Kenyon Brown, senior content acquisitions editor at Wiley, who provided suggestions on topics and content flow. It was out of these discussions that *In Silico Dreams* was formed; I thank Ken for his input at inception and guidance along the way. I also thank Gary Schwartz for expert editing and contributions to the flow of the book and to Ted Coombs for technical review and suggestions. I am also appreciative of the support of Christine O'Connor, managing editor; Saravanan Dakshinamurthy, Rachel Fogelberg, and the production staff; and Michael Trent for developing the cover design.

I gratefully acknowledge discussions, exchanges, and other communications that I had with individuals whose works and quotes are contained herein, including David Van Essen, Dan Felleman, Tushar Pandey, Philippe Schwabe, and Michael Wu for permission to use original figures, Emmanuelle Charpentier for clarifications on the potential of CRISPR technology, and many others in the STEM communities.

My intention in writing this book was to capture for the reader the intense pace of research and innovation occurring across many disciplines that are converging to impact medicine's future. Because of the enormous volume of work that has transpired over the past decade in artificial intelligence and biotechnology, it was impossible to cover all significant advances. I apologize for those omissions and for any errors in interpretation of technologies and research findings that I did cover as part of this project.

My personal journey inspired thoughts on many pages of this book. My meanderings across the once-separate fields of neuroscience, molecular biology, and computer science began in earnest while I was a freshman at the University of Washington in Seattle. I thank Donna Simmons for opening the door to neuroscience with a stint in her histology laboratory; Vijay Sarthy for allowing me to explore retinal neurochemistry and molecular biology in his research group; and Joel Levine for his patience and mentorship while I was a doctoral candidate in his lab, along with Simon Halegoua and many others at Stony Brook University. My first foray into industry was with James Morgan at the Roche Institute of Molecular Biology, a part of the pharmaceutical giant Hoffmann-La Roche in Nutley, New Jersey. I thank Jim for allowing me to plunge headlong into molecular biology.

I discovered the nascent power of combining computational approaches and software with biotechnology while working with Karl Hasel and Gregor Sutcliffe at Digital Gene Technologies in La Jolla, California. I thank Chuck for the opportunity and Greg for being a mentor and partner over the past two decades. To Graham Gaylard and the team at Real Time Genomics that provided me with a speed course in Reverend Bayes' theorem and machine learning, thank you.

I was supported in so many ways by my incredible wife Melanie Greenberg and daughter, Sydney, who both gave me the space and encouragement to complete this book. To Melanie and Sydney, thank you for your kindness and love.

—Brian Hilbush

Contents at a Glance

Contents

Introduction

We have entered an unprecedented era of rapid technological change where developments in fields such as computer science, artificial intelligence (AI), genetic engineering, neuroscience, and robotics will direct the future of medicine. In the past decade, research organizations around the globe have made spectacular advances in AI, particularly for computer vision, natural language processing and speech recognition. The adoption of AI across business is being driven by the world's largest technology companies. Amazon, Google, and Microsoft offer vast, scalable cloud computing resources to train AI systems and platforms on which to build businesses. They also possess the talent, resources, and financial incentives to accelerate AI breakthroughs into medicine. These tech giants, including Apple, are executing on corporate strategies and product roadmaps that take them directly to the heart of healthcare. Every few weeks, a new AI tool is announced that performs a medical diagnostic procedure at human levels of performance. The pace of innovation in the tech sector is exponential, made possible by continual improvements and widespread availability of computing power, algorithmic design, and billions of lines of software programming code. Technology's influence on the sciences has been profound. Traditional disciplines such as biology and chemistry are being transformed by AI and data science to such an extent that new experimental paradigms have emerged for research and the pharmaceutical industry.

Biotechnology's growth and innovation cycles are equally impressive. Startling advances have been made to move the field from simple gene cloning experiments using viral and bacterial genetic material in test tubes to performing gene editing at precise locations in the human genome. A new generation of gene therapy and T-cell engineering companies are building tools to equip the immune systems of patients to destroy cancer. Explosive growth in data-generating capabilities from DNA sequencing instruments, medical imaging, and high-resolution microscopy has created a perfect storm of opportunities for AI and machine learning to analyze the data and produce biological insights. Out of this milieu, the first generation of tech-inspired startups has emerged, initiating the convergence of AI and biotechnology. These young companies are taking aim at the conventional path of drug development, with the brightest minds and freshest ideas from both fields providing a new base of innovation for the pharmaceutical industry.

This book tells the story of the impact of innovations in biology and computer science on the future of medicine. The creation of a new industry based on therapeutic engineering has begun. Nearly 200 years ago, Emmanuel Merck saw a commercial opportunity to produce the pain-killing substance from the opium poppy, which was in widespread use across Europe and beyond. He was inspired by Fredrich Sertürner's innovative process for the extraction of the opiate alkaloid. Sertürner gave the newly purified narcotic substance the name morphium, after the Greek god of dreams. For thousands of years before these Germans helped to launch the pharmaceutical industry, medicinal compounds derived from nature had been concocted into noxious mixtures of uncertain potency by alchemists, physicians, or shamans in all cultures. With the elucidation of the rules of organic chemistry, the preparation and manufacturing of small molecule drugs and the practice of medicine would be forever changed.

The pharmaceutical industry began during the Industrial Revolution, drawing on a series of innovations in chemistry from the coal tar-based dye industry, along with other technological developments. This same rhythm of explosive innovation occurred again 100 years later in post–World War II laboratories in the United States and Britain. In the epochal years of 1952 and 1953, the foundations of computing, molecular biology, neuroscience, AI, and modern medicine arose almost at once, appearing in juxtaposition against the afterglow of the first thermonuclear bomb detonated in the Pacific. Science was literally blazing on all fronts.

Medicine has benefited enormously from the scientific discoveries and technologies born in the atomic age. Biotechnology has its roots in the principles and successes of molecular biology. The historic beginning was the discovery of the double helical structure of DNA in 1953, followed a generation later by the development of recombinant DNA technology in the 1970s. Therapeutics originating from biotechnology innovations now account for 7 of the top 10 drugs sold in the world.

Cancer chemotherapy treatments entered into clinical practice in the early 1950s, landmarked by the FDA's approval of methotrexate in 1953. These therapies provided a rational basis for attacking cancer cells selectively and sparked a decades-long search for new chemotherapeutics. As importantly, clinicians became critical in the evaluation of these and other new drugs in clinical trials, taking a seat at the table alongside medicinal chemists and pharmacologists as decision-makers in industry.

In neuroscience, Alan Hodgkin and Andrew Huxley's unifying theory of how neurons fire action potentials was published in 1952. The Hodgkin-Huxley model stands as one of biology's most successful quantitative models, elegantly tying together experimental and theoretical work. The framework led to the search for the ion channels, receptors, and transporters that control ionic conductance and synaptic activity, which together formed the basis of 50 years' worth of neuroscience drug discovery.

Modern computing and AI began with the work of its seminal figures starting in the 1930s and was anchored by successful operation of the first stored program, electronic digital computer—the MANIAC I—in 1952. Historian George Dyson framed the significance of this moment well in his 2012 book, *Turing's Cathedral: The Origins of the Digital Universe*, (Vintage, 2012), stating that "The stored-program computer conceived by Alan Turing and delivered by John von Neumann broke the distinction between numbers that mean things and numbers that do things. The universe would never be the same." AI pioneers who had hopes for machine intelligence based on neural networks would need another 60 years and a trillion-fold increase in computing performance to have their dreams realized.

The science and technologies sparking the biotech and digital revolutions developed in parallel over the past 50 years and within the past decade have acquired powerful capabilities with widespread applications. The convergence of these technologies into a new science will have a profound impact on the development of diagnostics and medicines and nonpharmaceutical interventions for chronic diseases and mental health. The recent advances in AI and biotechnology together will be capable of disrupting the long-standing pharmaceutical industry model via superiority in prediction, precision, theory testing, and efficiency across critical phases of drug development. Not too far into the future, with any luck, the *in silico* dreams of scientists and its impact on medicine will be realized.

What Does This Book Cover?

The book ties together historical background with the latest cutting-edge research from the fields of biotechnology and AI, focusing on important innovations affecting medicine. Several chapters also contain highlights of the crop of new businesses engaged in the latest gene and cell therapy, along with those founded on AI-based therapeutic discovery and engineering. An in-depth look at the history of medicines sets the stage for understanding the pharmaceutical industry today and the evolution of therapeutic discovery for tomorrow.

Chapter 1, "The Information Revolution's Impact on Biology," begins with an overview of milestones in technology innovation that are central to modern biology and biomedical applications. The first section covers the success of genomics in tackling the deluge of genome sequencing information during the COVID-19 pandemic and biotech's utilization of the data for creating a vaccine against SARS-CoV-2. The next section details the recent paradigm shift in biology, describing how the field is moving toward a more quantitative discipline. Another major thrust of the chapter is the role of computational biology in human genome sequencing, and its potential for medicine in the 21st century.

Chapter 2, "A New Era of Artificial Intelligence," covers the history of AI's development and the major milestones leading up to the stunning advances in deep learning. The role of neuroscience in formulating some of the ideas around artificial neural networks and the neurobiological basis of vision are discussed. An introduction to various approaches in machine learning is presented along with current deep learning breakthroughs. A first look at AI applications in medicine is also given. The chapter ends with a brief look at current limitations of AI.

Chapter 3, "The Long Road to New Medicines," travels all the way back to the Stone Age to reveal humanity's first random experimentations to find nature's medicines. The first section outlines the progression of therapeutic discovery through four eras: botanicals, chemical therapeutics, biotherapeutics, and therapeutic engineering. The next section delves into the industrial manufacturing of medicines and the rise of the modern pharmaceutical industry. The chapter describes the birth of chemotherapeutic drugs and antibiotics and the impact of war on their development. A segment is devoted to the development of cancer therapeutics, including immunotherapy. The latter sections cover the pharmaceutical business model of the 21st century and the role of biotechnology in drug discovery innovation.

Chapter 4, "Gene Editing and the New Tools of Biotechnology," begins by introducing the timeline and brief history of the development of precision genome engineering tools. A significant portion of the chapter covers molecular biology and biological information flow, with a history of recombinant DNA technology. The second-generation biotechnology tools from the bacterial CRISPR-Cas systems are outlined and presented as important genome editing strategies. A companion section reviews clinical trials of CRISPR-Cas engineered therapies. A final section describes the mRNA vaccine platforms and innovations leading up to its success against the SARS-CoV-2 virus.

Chapter 5, "Healthcare and the Entrance of the Technology Titans," provides a look at how each of the technology giants—Amazon, Apple, Google, and Microsoft—are making moves to enter the healthcare sector. The first section describes digital health and investment activity in this newly emerging area, along with the drivers of healthcare technology innovation. A series of vignettes presents the ability of each tech giant to disrupt and play a role as new participants in healthcare, with a look at their competitive advantages in the healthcare landscape.

Chapter 6, "AI-Based Algorithms in Biology and Medicine," explores how AI technology is already impacting biomedical research and medicine today and potential routes for the future. Two sections provide in-depth coverage of deep learning algorithms for cancer and brain diseases. The final sections review regulatory approval of AI-based software as a medical device and the challenges faced in implementation of clinical AI.

Chapter 7, "AI in Drug Discovery and Development," dives into the use of AI and machine learning in drug discovery. A brief survey of *in silico* methods in drug discovery and development is presented, followed by a section on computational drug design with AI tools. A subsequent section introduces biotechnology companies that are creating a new base of innovation for the industry. A final section summarizes where AI is deployed currently across pharmaceutical discovery and development.

Chapter 8, "Biotechnology, AI, and Medicine's Future," begins with a discussion of convergence and how a new discovery engine based on hypothesis generation and evaluation by AI might work across biology, pharma, and medicine. The next section looks at how experimental approaches and computational methods together power biology by forming a new tech stack. AI's potential for neuroscience and the value of brain studies for AI and medicine are presented around the theme of motor control behavior and the brain. The chapter ends with a look at the landscape of companies arrayed against the range of technologies being developed to engineer therapeutics.

Reader Support for This Book

If you believe you've found a mistake in this book, please bring it to our attention. At John Wiley & Sons, we understand how important it is to provide our customers with accurate content, but even with our best efforts an error may occur.

To submit your possible errata, please email it to our Customer Service Team at `wileysupport@wiley.com` with the subject line "Possible Book Errata Submission."

The Information Revolution's Impact on Biology

I think the biggest innovations of the twenty-first century will be at the intersection of biology and technology. A new era is beginning, just like the digital one..."

Steve Jobs in an interview with Walter Isaacson

The transformative power of the information revolution has reverberated across all industry sectors and has profoundly altered economies, political landscapes, and societies worldwide. Among the scientific disciplines, physics, astronomy, and atmospheric sciences were the first to benefit directly from the development of mainframe computing and supercomputers from the 1960s onward. The most dramatic advances came after the development of semiconductor electronics, the personal computer, and the Internet, which further accelerated the information revolution. These historic innovations were the catalysts for producing amazing new technological capabilities for biological sciences and for the biotechnology and pharmaceutical industries.

Scientific progress in biology is highly dependent on the introduction of new technologies and instrumentation with ever-increasing resolution, precision, and data gathering capacities and output. Table 1.1 breaks down the milestones in technology innovation by decade.

For much of the twentieth century, biology borrowed equipment from physics to visualize cellular and macromolecular structures and measure atomic-scale dimensions. In the pre-digital era, the observations and experimental data of scientists were captured on paper and physical media such as magnetic drums, tapes, X-ray films, and photographs.

Table 1.1: Milestones in Technology Innovation by Decade

DECADE(S)	TECHNOLOGY	FIELD	APPLICATIONS	REFERENCE
1900s–1930s	Electron microscopy	Physics	Structural biology	Chapter 1
	X-ray crystallography	Physics	Structural biology	Chapter 1
	Electrocardiogram	Physics	Cardiology	Chapter 6
1940s	Broad spectrum antibiotics	Microbiology	Medicine	Chapter 3
	NMR	Physics	Biology, chemistry, drug discovery	Chapter 4
1950s	Confocal microscopy	Physics	Biology	Chapter 7
	Artificial intelligence	Computer science	Information technology	Chapter 2
1960s	Laser	Physics	Information technology and instrumentation	Chapter 1
	Ultrasound	Physics	Medicine	Chapter 7
	Semiconductor electronics	Tech industry	Information technology and instrumentation	Chapter 2
	Integrated circuits	Tech industry	Information technology and instrumentation	Chapter 2
1970s	MRI, PET, CT	Physics and chemistry	Medical imaging and biology	Chapter 6
	Recombinant DNA	Molecular biology	Biology, drug discovery, therapeutics	Chapter 4
	Single-channel recording	Biophysics	Biology and drug discovery	Chapter 3
	Monoclonal antibodies	Immunology	Drug discovery and diagnostics	Chapter 3
	RAM memory	Tech industry	Information technology and instrumentation	Chapter 1

DECADE(S)	TECHNOLOGY	FIELD	APPLICATIONS	REFERENCE
1980s	PCR	Molecular biology	Biology, drug discovery, diagnostics	Chapter 1
	DNA sequencing	Molecular biology	Biology, drug discovery, diagnostics	Chapter 1
	DNA synthesis	Chemistry	Biology and drug discovery	Chapter 4
	MALDI-TOF/ESI MS	Biophysics	Biology and drug discovery	Chapter 3
	Personal computing	Tech industry	Information technology	Chapter 1
1990s	fMRI	Physics	Medical imaging	Chapter 5
	Two-photon microscopy	Physics	Biology	Chapter 4
	Transgenic technology	Molecular biology	Biology and drug discovery	Chapter 3
	RNAi	Molecular biology	Biology, drug discovery, therapeutics	Chapter 1
	Internet	Tech industry	Information technology	Chapter 1
2000s	CAR-T	Immunology	Therapeutics	Chapter 4
	Stem cell reprogramming	Immunology/ hematology	Biology, drug discovery, therapeutics	Chapter 4
	Next-generation sequencing	Molecular biology	Biology, drug discovery, diagnostics	Chapter 1
	Optogenetics	Biophysics	Biology	Chapter 3
	Cloud computing	Tech industry	Information technology	Chapter 2

Continues

Table 1.1 (*continued*)

DECADE(S)	TECHNOLOGY	FIELD	APPLICATIONS	REFERENCE
2010s	Cryo-EM	Biophysics	Drug discovery	Chapter 1
	CRISPR gene editing	Molecular biology	Gene therapy, drug discovery, diagnostics	Chapter 4
	Single-cell sequencing	Molecular biology	Biology and drug discovery	Chapter 8
	Deep Learning	Computer science	Biology, drug discovery, diagnostics	Chapter 2
	Quantum computing	Tech industry	Information technology	Chapter 8

The advent of microprocessor-based computing brought about the realization of analog-to-digital data conversion and, along with that, random access memory (RAM) on semiconductor circuits. Digitization of data streams and availability of petabyte-scale data storage have been immensely important for conducting modern science, not only allowing researchers to keep pace with the deluge of information, but also enabling network science and the widespread sharing of research data, a fundamental feature of scientific progress.

The information revolution's impact on biology continues unabated into the twenty-first century, providing computing power that continues to grow exponentially, producing sophisticated software for data acquisition, analysis, and visualization, and delivering data communication at speed and scale. New disciplines have been launched during this era in large part due to the introduction of technologies and instruments combining computation with high resolution. Some of the most important are DNA synthesizers and sequencers, which led to genomics and computational biology, fMRI for computational neuroscience, cryo electron microscopy (cryo-EM), NMR, and super-resolution microscopy for structural biology and several compute-intensive spectroscopic techniques (for instance, MS/ MALDI-TOF, surface plasmon resonance, and high-performance computing) that opened up computational drug discovery. Medicine has similarly advanced in the twentieth century by the application of computational approaches and breakthroughs in physics that were combined to create an array of imaging technologies.

As a consequence of transporting biology from a "data poor" to a "data rich" science, the information revolution has delivered its most fundamental and unexpected impact: causing a paradigm shift that has turned biology into a quantitative science. Biological science and biomedical research are now benefiting from the tools of data science, mathematics, and engineering, which had been introduced in "big science" endeavors, that is, projects such as the Human Genome, Proteome, and Microbiome Projects[1-3]; the Brain Initiative and Human Brain Project[4,5]; international consortia such as the Cancer Genome Atlas and International Cancer Research Consortium[6-7]; and precision medicine and government-backed population health projects like All of Us in the United States, the UK Biobank, and GenomeAsia 100K.[8-10] These projects propelled forward the development of the "omics" technologies, most importantly genomics, epigenomics, proteomics, and metabolomics, with new quantitative approaches imagined and inspired by the information revolution to manage and analyze "big data."

Chapter 1 of this book will explore the information revolution's impact on biology. Computing's massive influence on industry has also been referred to as the third industrial revolution. Subsequent chapters of the book will deal with aspects of the fourth industrial revolution,[11] described as the confluence of highly connected sensors comprising the Internet of Things (IoT), machine learning and artificial intelligence, biotechnology, and digital manufacturing that is creating the future. Over the next few decades, the technologies powering the fourth industrial revolution will bring about *in silico biology*. Similar to the economic transformations occurring in banking, manufacturing, retail, and the automotive industry, the pharmaceutical industry is poised to see enormous returns to scale by embracing the coming wave of innovations.

A Biological Data Avalanche at Warp Speed

The breathtaking speed with which the worldwide medical and scientific communities were able to tackle the coronavirus pandemic was a direct consequence of the information revolution. The Internet and wireless communication infrastructures enabled immense amounts of data from viral genome sequencing efforts and epidemiological data to be shared in real time around the world. Digital technologies were essential for gathering, integrating, and distributing public health information on a daily basis. In the private sector, computationally intensive drug discovery pipelines used artificial intelligence algorithms and biotechnology innovations to accelerate compound screening, preclinical testing, and clinical development. Nearly every government-backed initiative, industry partnership, and global collaborative research effort was powered by cloud-based computing resources. The mountain of data provided insights into the nature of the disease and inspired hope that treatments and effective countermeasures would arrive soon.

The biomedical research and drug development communities went into emergency action almost immediately, sensing that time was of the essence, but also spotting opportunities for business success and scientific achievement. Researchers in thousands of laboratories were formulating new therapeutic hypotheses based on incoming data from viral genome sequence, virus-host interactions, and healthcare systems.

During the initial phase of the pandemic, an unprecedented trove of knowledge became available via a torrent of publications. The rapid publishing of more than 50,000 documents announcing early research results on medRxiv and bioRxiv servers provided an invaluable platform for reviewing new studies on anything from clinical and biological investigations of viral replication to complex, multinational clinical trials testing an array of potential therapeutics and vaccines. Unfortunately, there was also an urgency to vet ideas, drugs, and important public health policy measures in real time. Many of these failed, were premature, or became conspiracy theory fodder. Technology can only do so much, and COVID-19 has shown us that science, too, has its limits. In addition, choices in the political sphere have enormous consequences on outcomes of scientific endeavors, public health, and the future of medicine.

The tragic irony of the SARS-CoV-2 outbreak is that China had an impressive defense to protect against a second SARS-type epidemic event, built around information technology. The Contagious Disease National Direct Reporting System[12] was engineered to facilitate reporting of local hospital cases of severe flu-like illnesses and deadly pathogens such as bubonic plague and to deliver notification to authorities within hours of their occurrence anywhere in China. Health officers in charge of nationwide disease surveillance in Beijing could then launch investigations, deploy experts from the China Centers for Disease Control, and set strategy for regional and local municipalities to deal with escalating public health situations. The design had a weak link that proved disastrous—hospital doctors and administration officials, fearing reprisals and hoping to contain the damage, decided not to use the alarm system.

On December 31, 2019, more than 200,000 Wuhan citizens and tourists participated in the New Year's Eve light show along the banks of the Yangtze River. While the world was poised to celebrate the arrival of a new decade, top Chinese health officials were about to learn the shocking news: another epidemic caused by a SARS-like coronavirus was underway, and Wuhan, a city of 11 million inhabitants, was at ground zero. Before the news rocketed around the globe, Chinese government officials, health authorities, and scientists had reviewed results of DNA sequence analysis that made the discovery possible. Another call was placed to the World Health Organization's (WHO) China Representative Office, announcing the laboratory results outside of the Chinese bureaucracy.[13,14] Perhaps no other technology was more critical at the outset of the pandemic than genomic sequencing.

Like other epidemics, the outbreak started in fog, but by late December, at least a half-dozen hospitals all across Wuhan were reporting cases of a severe flu-like illness that led to acute respiratory distress syndrome and pneumonia. The first confirmed case was retrospectively identified on December 8, 2019. Several patients were being transferred out of less well-equipped hospitals and clinics and placed into intensive care units within Wuhan's major hospitals, including Wuhan Central and Jinyintan Hospital, a facility that specialized in infectious disease. Physicians became alarmed as standard treatments were not effective in reducing fever and other symptoms.

As the situation worsened and clinical detective work was hitting a wall, a small handful of doctors began to see a pattern emerging from the incoming cases. Nearly all patients who shared similar clinical characteristics either worked at or lived nearby the Huanan seafood market, suggesting they had encountered a common infectious agent. Since preliminary laboratory tests had ruled out common causes of pneumonia due to known respiratory viruses or bacteria, doctors realized that a potentially new and highly contagious pathogen was circulating in the community. The cluster represented the beginnings of an epidemic: Jinyintan Hospital quickly became the focal point, seeing the first patients in early December; cases developed at Wuhan Red Cross Hospital (December 12), Xiehe Hospital (December 10–16), Wuhan City Central Hospital (December 16), Tongji Hospital (December 21), Hubei Hospital of Integrated Traditional Chinese and Western Medicine (December 26), and Fifth Hospital in Wuhan.[13–15]

On December 24, 2019, the pressure had become so intense within Wuhan City Hospital that Dr. Ai Fen, who headed the emergency department, rushed a fluid sample from a critically ill patient to Vision

Medicals, a gene sequencing laboratory in Guangzhou, 620 miles away (in hospital settings, the need for rapid and low-cost diagnostics usually means that any high-precision molecular diagnostic test is performed by an outside laboratory). Over the next few days, a collection of samples was sent to BGI, the powerhouse genomics institute, and another to CapitalBio MedLab, both in Beijing.[13,14] Through the lens of unbiased sequence analysis with metagenomics, any microorganism with nucleic acid (DNA or RNA) in the samples could be traced, including the possibility of discovering new species with similarities to existing ones.

Within three days, genome sequencing and bioinformatics methods had determined that the causative pathogen was a SARS-like coronavirus and a species distinct from SARS-CoV, which was responsible for the pandemic that also erupted in China and spread globally from 2002–2003. The pieces were rapidly coming together. SARS-CoV was initiated by zoonotic transmission of a novel coronavirus (likely from bats via palm civets) in food markets in Guangdong Province, China. Now, yet another coronavirus had likely crossed over to humans with deadly virulence in Wuhan's Huanan market. Although very little was known yet about the virus's features, the DNA sequence information of the new viral genome portended the genesis of a devastating pandemic that would reach into 2020 and beyond.

Wuhan hospital and local health officials were only beginning to sense the gathering storm. On December 27, 2019, the first molecular test results were relayed to hospital personnel by phone from Vision Medicals due to the sensitivity of the situation.[13] Another report mentioning a possible SARS coronavirus came in from CapitalBio. A doctor at the Hubei Hospital of Integrated Traditional Chinese and Western Medicine had also raised the alarm on December 27. Although the timing and sequence of events remains murky, epidemiological experts from the Hubei Provincial Center for Disease Control, the Center for Disease Control in Wuhan, and several district-level disease control centers were tasked sometime around December 29 to gather evidence and identify all potential cases across the city.

The likelihood that a new coronavirus was circulating in a highly contagious manner, potentially endangering staff, led to leaks on social media. Ai Fen, who had notified hospital leadership of results on December 29, decided to disseminate the test report. She circled the words *SARS coronavirus* in red and sent out a message to fellow staff.[14,15] The startling revelation was seen by Li Wenliang, an ophthalmologist at Wuhan Central, who was concerned enough to send out texts using WeChat in a group chat that reached 100 people, which amplified the alarm more broadly outside of Wuhan Central on December 30. It was a courageous move. The texts were intercepted by Chinese authorities, and a few days later, Dr. Li was sent an admonition notice by the Wuhan Municipal Public Health Bureau.[16]

In Beijing, the diagnostic test information from the hospital sources circulating on social media made its way to Gao Fu, the director of the China Center for Disease Control and Prevention. He was stunned. From his position at the pinnacle of the public health emergency response hierarchy in China, he bore the ultimate responsibility and was sharply critical of the Wuhan health officials. Why had the early cases not been reported through the system? Within hours, he began coordinating the CDC and National Health Commission's response to send clinicians, epidemiologists, virologists, and government officials to Wuhan. The epidemiological alert was officially announced on December 31, 2019. However, pressure to avoid political embarrassment and public alarm delayed full disclosure of the investigations until the second week of January 2020.

From news reporting and Chinese press announcements, it is unclear how convinced the various Chinese authorities were of the existence of a new SARS coronavirus or of its routes of transmission. On the one hand, they knew that the disease leading to fatal pneumonia had clinical characteristics highly similar to SARS. And the genomic sequencing from multiple patient samples, even if incomplete, strongly suggested a new SARS-like coronavirus. If true, an immediate need arose to get a diagnostic test established, which would be completely based on the viral genome sequence. The new virus would have to be isolated and cells inoculated in the laboratory to study the cellular pathology in detail. Thus, in parallel with the on-site epidemiological investigations, plans were made to obtain further DNA sequencing from top-tier laboratories, including the government's own facilities.

Remarkably, the rest of the world had its first alerts arrive on the same day that Li Wenliang notified his colleagues on WeChat. At least three separate systems picked up anomalies by AI algorithms. The HealthMap system at Boston Children's Hospital sent out an alert at 11:12 p.m. Eastern time, as it detected social media reports of unidentified pneumonia cases in Wuhan—but the alert went unnoticed for days. Similar warnings were issued by BlueDot, a Toronto-based AI startup and Metabiota in San Francisco.[17,18]

Back in Wuhan, the race to sequence the novel coronavirus genome had begun. On the morning of December 30, 2019, a small medical van sped away from Wuhan's Jinyintan Hospital and crossed over the Yangtze river on its way to the Wuhan Institute of Virology. The driver was transporting vials that contained lower respiratory tract specimens and throat swabs taken from seven patients previously admitted to the intensive care unit (ICU) of the hospital with severe pneumonia. The institute, home of China's highest security level biosafety research, set a team of scientists in the Laboratory of Special Pathogens to begin work on the incoming samples. In two days, they generated confirmatory results and had fully mapped genomes from five of the seven samples. The analysis showed that the virus present in each of the sequenced samples had essentially the same genome; a new virus had entered the city and was likely being transmitted by human contact.[19,20]

Another research team, 1,000 miles away in Shanghai and led by Yong-Zhen Zhang at Fudan University and Shanghai Public Health, received samples by rail on January 2, 2020, and had the viral genome sequencing results within three days. On January 5, the Shanghai-based researchers notified the National Health Commission that they had derived a complete map of a new SARS-like coronavirus.[21] Not to be outdone, a China CDC laboratory in Beijing finished another genome assembly with a trio of sequencing methods by January 3, and it was the first group to publish results in the CDC weekly on January 20.[22] More complete analysis derived from the metagenomic and clinical sequencing efforts followed in high-impact journals.

It became clear that by the opening days of 2020, the Chinese government had in its possession all of the evidence it needed to declare an outbreak. DNA sequencing technology had provided comprehensive molecular genetic data proving the existence of a novel virus. Emerging epidemiological data on dozens of cases since mid-December indicated contagious spread, although incontrovertible proof of human-to-human transmission apart from the Huanan market exposure was lacking. The confusion on the latter may have led the government to decide to conceal or withhold the information. As time moved on, it looked more and more like a coverup. Why the delay? The most likely explanation is that there still was some hope that a local containment strategy in Wuhan could extinguish the spread. Could authorities stave off an epidemic before the virus engulfed China and spread globally? At Communist Party Headquarters, there were likely considerable internal deliberations, lasting days, as to what the country's response measures would be to control a highly contagious virus. The leadership was confronting the reality that an uncontrolled outbreak would wreak havoc on a population with no immunity to a deadly new virus.

President Xi ordered officials to control the local outbreak on January 7, 2020. It was not until January 9, a week after the sequencing was done, that the Chinese health authorities and the World Health Organization (WHO) announced the discovery of the novel coronavirus, then known as *2019-nCoV*. The sequencing information was finally made publicly available January 10 (`https://virological.org/`).[15] Wuhan was ordered to be under lockdown starting January 23, followed the next day by restrictive measures in 15 more cities in Hubei Province, effectively quarantining 50 million people. China braced for major disruptions heading into the Lunar New Year celebrations. For the rest of the world, however, critical weeks had been lost to prepare for and contain the epidemic.

The release of the genome data turned out to be vital in the early stages of the pandemic in 2020 for tracking the virus and, perhaps more importantly, for building a scientific plan for genome-based diagnostics, vaccines, and drug development strategies. Through the jumble of hundreds of millions of nucleotides (the familiar A, C, G, and Ts standing for the chemical bases that are the building blocks of DNA) produced on sequencing instruments and turned into bytes, computational biologists used software to construct an approximately 30,000 nucleotide long "assembly" containing the entire genome of the virus. From there, researchers demarcated the boundaries and "mapped" every gene contained along the linear extent of the genome (see the section "Analyzing Human Genome Sequence Information").

The ability to determine the identity of the sequence, and indeed to assemble the fragments, relies on comparing the sequence fragments back to known genomes contained in biological sequence databases. Algorithms designed to find exact or highly similar matches of a short sequence from a swab or fluid sample to those contained in the database provide the first glimpse of what organism or organisms are collectively found in the biological material. For characterizing and classifying the virus genome, the answers come by comparing a reassembled DNA sequence to a universal database containing complete genomes of tens of thousands of known bacterial species, thousands of viruses, and an array of other exotic, pathogenic genetic sequences.

Tracking SARS-CoV-2 with Genomic Epidemiology

What is gained from the genomic sequence information goes way beyond the initial determination of a pathogen's identity and extends into four main areas: epidemiology, diagnostics, vaccines, and therapeutics (antivirals and other drug modalities). For decades, virologists and epidemiologists, together with public health systems, have been able to acquire and analyze pathogen sequences, but only recently have newer sequencing technologies, known as *next-generation sequencing* (NGS) and *third-generation sequencing* (nanopore or single-molecule sequencing), enabled genomic results from infectious cases to be obtained in days versus months. An entirely new and data-rich method for tracing outbreaks worldwide arose with genomics-based epidemiological methods.

The basis of genomic epidemiology is that the sequencing readout can detect changes at single nucleotide resolution using a reference genome. At every position of a genome's sequence, differences can be spotted by comparing to other genomes, say within a family or community, or to entire populations. The key to this approach is that viruses (RNA viruses in particular) leave telltale molecular tracks as they replicate in an error-prone way in their host cells, leading to mutations that can be detected later. Once new viral particles are shed out into the environment and then infect another host, the replication cycle starts again. In the case of SARS-CoV-2, copies of its RNA genome are made with an enzyme known as *RNA-dependent RNA polymerase* (the target of the antiviral drug remdesivir). Viral replication mistakes, or mutations, produced by this viral enzyme are acquired and passed on when copying the genome. These mutations can be "neutral" and not affect viral protein function or the ability of the virus to survive. However, new mutations might introduce some survival advantage by random chance and be propagated. Deleterious mutations that disable virus replication or other essential features are dead-ends and lost, while others remain to potentially hasten a virus' disappearance. Tracking these changes over time takes on the appearance of a family tree; building these genetic ancestry trees using sequencing information is genomic epidemiology. The mutations that pile up in the genome over time occur at random within the genome, and thus different tree branches will have a unique set of mutations to label the branch. This type of high precision tracing information is unique to genome sequencing data.

The first use of high-throughput sequencing for genomic surveillance came during the Ebola virus epidemic of 2013–2016.[23] It was a landmark achievement in epidemiology. For the first time in history, genome surveillance could be used as a tool to investigate transmission routes and provide intelligence for outbreak responses. In addition, a pathogen's evolution could be tracked as the epidemic proceeded. The Ebola experience led the scientific community to build an information technology infrastructure for storing and sharing genomes and to provide computational tools for genomic epidemiological analysis. One of the first efforts,[24] known as the Global Initiative on Sharing All Influenza Data (GISAID), was initially developed to share sequences and monitor the genetic evolution of influenza viruses. Another effort relevant to the coronavirus pandemic was the nextstrain.org open source platform for pathogen surveillance, built by a team led by Trevor Bedford of the Fred Hutchinson Cancer Research Center, Seattle, and by Richard Neher of Biozentrum at the University of Basel, Switzerland.[25] In 2017, the group won the Open Science Prize to complete the platform. The finished product came just in time.

Once the coronavirus sequence was available in January 2020, the digitally empowered global resources went into action. As the virus spread out of China, sequences of the virus isolated from Asia, Europe, and the United Kingdom, North America, and other countries were submitted to GISAID. Bedford and his colleagues took the incoming genomes and began reconstructing the movement of the virus as it spread across the globe. Researchers elsewhere were also tracing the spread and using the genomic data to determine whether strain differences were appearing. Genomic evidence compiled during the early months of the outbreak revealed important features of the virus. First, very few mutations were accumulating, indicating that the virus had only recently infected humans. The new coronavirus was changing at a slower rate relative to other viruses, with less dynamic branching. The structure of the nascent tree, shown in Figure 1.1, reveals that the trunk was anchored to a single source and indicates that the virus descended from a single case in Wuhan province.

At the outset of the global spread, genome tracing capabilities were not yet in place, and hunting down where transmissions were occurring was difficult. As an example, the genomes derived from diagnostic patient samples in Canada, the United Kingdom, and Australia were identical, or nearly identical, suggesting some sort of linkage. After a review of some of these patients' social activities, epidemiologists noted that all of these early cases shared one thing in common: recent travel to Iran. Without the genomes, it is unlikely that such a connection would have even been noticed. The genome surveillance also provided the first evidence in the United States that community spread was occurring in Seattle.[26] An individual identified as the first case that had traveled back from Wuhan had a nearly identical genomic signature to another COVID-19 case in the region. The hope with the genomic epidemiological tools was that close monitoring of spread, even with a small number of genomes, could aid public health officials in relaxing the burdensome social-distancing measures that were important in controlling the pandemic. At the most basic level, the genomic data could provide the authorities with information as to whether the new infections were introduced from travelers or were from local spread. Unfortunately, with SARS-CoV-2, the slower mutation rate does not allow for accurate inference of transmission change in same way that worked for HIV, where a unique viral genotype appeared after essentially every transmission.

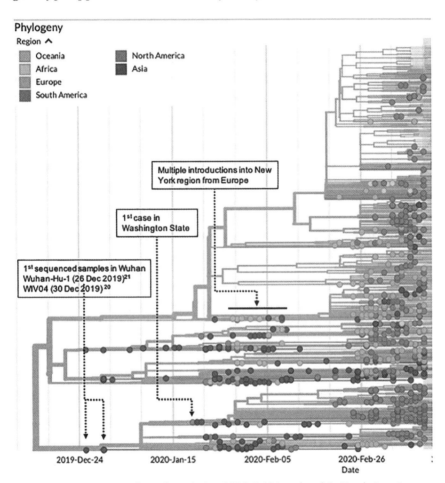

Figure 1.1: Genomic epidemiology during SARS-CoV-2 outbreak in North America

Phylogenetic tree showing relationship of 828 genomes sequenced from SARS-CoV-2 infected individuals between December 2019 and March 2020. Dots represent one individual labeled by geographic origin, as in the legend. The tree is derived by comparing viral genome sequence data between individuals, where mutations create a new branch and vertical distance is indicative of degree of difference between genomes. Individuals on same horizontal line share virus with identical genomes, which enables tracing back to a common ancestor .*Source*: Nextstrain.org under CC-BY-4.0 license.

At no other point in history had the world detected an outbreak of infectious disease and deciphered the genome of the responsible pathogen in a matter of weeks. It is rather hard to imagine in the digital age just how slowly and methodically biologists and clinicians had to proceed to work out the etiology of mysterious diseases, such as AIDS. Progress in discovering HIV as the cause of AIDS was dauntingly slow, given the complexity of the disease course and the numerous technical, medical, and societal hurdles that existed at the time. Three years of intensive research were needed from the first formal identification of an AIDS patient in the United States in 1981 to the discovery of the new HIV virus (originally named HTLV-III) in 1984.[27] Fortunately, the tools of molecular biology had been developed in that time frame, allowing genes to be isolated and cloned and later sequenced by manual methods (see Chapter 4, "Gene Editing and the New Tools of Biotechnology"). Other important technical advances were needed, such as the development of a cell culture technique to propagate the virus and animal models to study disease progression. The most significant breakthrough was the identification by research clinicians of a unifying biological marker of disease, a significant loss of a population T cells (CD4+). With AIDS/HIV, the determination of the retrovirus genome sequence played a minor role in understanding the disease and was completed only toward the end of the discovery phase.

The AIDS pandemic set into motion a much more focused awareness of zoonotic viruses and the threat that such pathogens present to the world. Prior to AIDS, most in the medical community had viewed exotic viruses and spread of plague due to *Yersinia pestis* from fleas carried by rodents as no longer threats to industrialized societies. In the United States, the last case of smallpox was recorded in 1949. Principally through worldwide vaccination programs, smallpox was eradicated by the late 1970s. If you were to inspect a list of the top five infectious diseases in the United States in 1979, you would find chickenpox (199,081 cases), salmonellosis (33,138), Hepatitis A (30,407), syphilis (24,874), and Hepatitis B (15,452).[28] Effective vaccines were available and had drastically reduced the incidence of many dangerous "childhood" diseases, including measles, mumps, rubella, and poliovirus. Yet new animal-to-human transmissions continued to occur and not respect boundaries, whether the virus outbreaks began in Africa, Asia, or elsewhere. The names are now scary and familiar: Hantavirus (1993, Southwestern United States), West Nile Virus (1996, Romania; 2002 United States), Zika virus (multiple epidemics: 2007, 2013, 2015, 2016), and Ebola (Guinea, 2013–2016). By 2018, the infectious disease landscape in the United States had changed dramatically. On the CDC's list of the most concerning pathogens were those having animal reservoirs, with influenzas (avian and swine), coronaviruses (bats, camels, birds), and West Nile Virus (mosquitoes) in the top five.[29]

Emerging pathogen detection had advanced beyond simple polymerase chain reaction (PCR) and conventional virologic methods at the start of the twenty-first century. High throughput sequencing instruments were in hand by the time the first coronavirus epidemic began with the emergence of severe acute respiratory syndrome (SARS) in 2002–2003. Once the virus was isolated and culture methods could produce sufficient quantities of viral particles to extract genetic material, automated sequencers combined with gene assembly algorithms and pieced together the novel coronavirus genome in five months (November 2002 to mid-April 2003; the technical sequencing steps required only 31 days).[30] Fast-forward 10 years to 2012, and another coronavirus was found, having leapt from camels to humans. Sequencing technology advanced exponentially during this time period with next-generation sequencing, which determined the genome of *Middle Eastern Respiratory Virus (MERS)*, a distant SARS relative, within the space of three months.[31] The availability of the sequence information and molecular diagnostic tools allowed for actions to be taken before the virus could cause a more widespread epidemic.

DNA sequencing technology played a role only after a few months following the Ebola outbreak in a remote village in Guinea in December 2012.[32] After the outbreak was detected, it took three weeks to produce genome sequences from patient samples. Although genomic epidemiology was used in real time to characterize viral spread by survivors and by sexual contact, much more could have been done with the data had sequencing rolled out earlier. Finally, with SARS-CoV-2, the timespan from unknown cause to confirmed identity in precise molecular detail had shrunk to three days. Although much more was needed to be learned and understood about the virus, genome sequencing and computational methods delivered the essential goods to begin to tackle the pandemic. Uncovering the data embedded in the viral genetic material, the RNA molecule spanning 29,123 nucleotides, would unleash a massive, worldwide effort in the hunt for life-saving drugs and vaccines.

Biology's Paradigm Shift Enables *In Silico* Biology

"Those who think 'Science is Measurement' should search Darwin's works for numbers and equations."
— David H. Hubel in: *The History of Neuroscience in Autobiography*
"An experiment is a question which science poses to Nature and a measurement is the recording of Nature's answer."
— Max Planck in: *Scientific Autobiography and Other Papers*

It is an understatement to say that biology and physics have long maintained distinct relationships with theory and mathematics. Although both fields rely heavily on experimental data and observation, the complexity of biological phenomena has evaded formulation by equations that capture fundamental principles. Ever since Galileo and the scientific revolution, physics has been spectacularly successful in building predictive frameworks for understanding natural laws in precise, quantitative terms. Think Maxwell's equations, Einstein's $E = mc^2$, or Newton's laws of motion. Is it possible that at some point biology will become a quantitative science with a series of equations that can make predictions from any biological process?

Biologists, and most scientists up until 30 years ago, would undoubtedly have answered that question with a resounding no. Evolutionary theory has unified all of biology and was built solely on observations made by Charles Darwin and Alfred Russel Wallace. Gregor Mendel's laws of heredity were a step closer; he arrived at his conclusions using a quantitative, experimental approach. The closest genetics has to a rigorous formulation is the underlying principle of population genetics, the Hardy-Weinberg law of equilibrium. The law states that the frequencies of both alleles and genotypes will remain constant from generation to generation in large populations that are not evolving. Hardy-Weinberg equations can be used to infer genotype frequencies when allelic frequencies are known and certain conditions are met. Deviations from the equilibrium are a measure of genetic variation.

For the past 70 years, the principles of genetics have been studied within the powerful framework of molecular biology. Mechanistic details of DNA replication, RNA transcription, and protein translation—biology's information processing system—are now well understood. What is not clear is whether algorithms or equations could calculate or describe precisely how gene regulatory networks act to control extremely intricate cellular processes, build nervous systems, orchestrate the development of organisms, or drive the evolutionary origin of species. Universal laws of growth, applied to biological systems, have been postulated and debated for nearly a century since Kleiber's first formulation of a scaling law for metabolic rate.[33–36] The observation that nature's living organisms obey many power scaling laws lends hope that new theories of biological control and processing can be formulated and tested in rigorous ways and provide scientists with predictive frameworks toward understanding biological phenomena.

The view that biological systems are intractable in mathematical terms may start to fade as tools to generate exponentially greater volumes of experimental data and new computational approaches provide power for the science of complexity in general. Steven Hawking was asked in the year 2000 if the twenty-first century would be the golden era of biology like physics was in the twentieth century, and his answer was that "the next century would be the century of complexity."[37] The tools to address complexity are what the field needs to move biology into the realm of a truly quantitative science.

The paradigm shift in biology has already begun as a result of the information revolution and the parallel introduction of technologies that produce data at scale for biological discovery. As an example, the output of a modern DNA sequencer, such as the HiSeq 4000 built by the San Diego genomics company Illumina, requires complex chemistry and molecular cloning steps in addition to a final stage that captures high-resolution pictures of millions of "sequencing by synthesis" reactions. Its output is a prodigious 1.5 terabytes of data in 1.5 days, sufficient for full sequencing of six human genomes, each with 3×10^9 bases of DNA.[38] Once the computational analysis pipeline produces an assembled genome, the genetic sequence provides a precise, *in silico* template, which serves as a starting point for cancer genome investigations, building drugs, or designing vaccines.

Transitions and Computation in Cancer Research

Comparing the progress made in cancer research from pre- and post-computing eras reveals the startling breadth of the paradigm shift. Starting in the 1970s, the discovery of the first oncogenes was accomplished at laboratory benches by individual scientists performing experiments with viruses and using cell culture methods. In a series of classic molecular genetic experiments completed in 1970, Peter Duesberg and Peter Vogt identified the transforming DNA of the first oncogene (called *src* for sarcoma) from avian Rous sarcoma virus (RSV).[39] While inconceivable today, determination of its DNA sequence and encoded protein product would have to wait another decade for the development of recombinant DNA technology (molecular cloning) and methods for sequencing DNA. Molecular cloning techniques had been developed and became available only in the mid-1970s (see Chapter 4), which gave researchers the ability to shuttle DNA containing foreign genes from one source into bacterial and mammalian cells.

To understand the biochemistry of the products of cancer genes such as *src* and the mystery behind cancer in general, cumbersome methods were needed to isolate and study the proteins. For the *src* gene product, the first successful isolation was accomplished by Joan Brugge and Ray Erickson in 1977 using antisera from immunized rabbits to capture the protein.[40] Subsequent biochemical studies by Erickson's group in Denver[41]; J. Michael Bishop, Harold Varmus, and colleagues at the University of California, San Francisco[42]; and Tony Hunter[43] at the Salk Institute led to the discovery that *src* encoded a protein tyrosine kinase.

The seminal discovery around the viral oncogene's origins came from Bishop and Varmus. They hypothesized that viral oncogenes might have been acquired from normal cellular genes involved in cellular growth control. Was it possible to find such a *src* relative (a cellular homolog) in humans and other species? What would lead the adopted gene to become oncogenic when carried by viruses? Using radioactively labeled DNA probes and molecular hybridization assays, they were able to demonstrate that several avian genomes contained *src*-like genes based on their ability to complex, or form hybrids, with the viral DNA (*note*: RSV has a RNA genome, so the *src* RNA was converted to a DNA copy using a retroviral enzyme known as *RNA-dependent DNA polymerase*, also known as *reverse transcriptase*).[44]

Once DNA sequencing methods became available in the early 1980s, the *src* genes from viruses (RSV and cousin ASV) and avian species, as well as humans, were sequenced.[45,46] The analysis of the DNA sequence proved that the viral oncogene originated from a cellular "proto-oncogene." By the time Bishop and Varmus had won the Nobel Prize for their work on retroviruses and oncogenes in 1989, at least 60 other proto-oncogenes had been identified through molecular techniques and DNA sequencing. The normal functions of most of these cellular proteins, including *src*, were in signaling pathways and regulatory circuits controlling cellular growth and differentiation. Examination of DNA gene sequences revealed that cancer was indeed a genetic disease, caused by mutations within the genes. The types of mutations causing genetic "lesions" could be seen broadly via DNA sequencing as changing the amino acid sequence of the encoded protein by addition or subtraction of DNA bases (insertions or deletions) and by single nucleotide changes referred to as *point mutations* (also by the shorthand SNP or SNV, for single nucleotide polymorphism or variant, respectively). For *src*, human cancers were not found to have mutated forms of the gene; rather, extra copies are present that lead to gene amplification and overexpression of the signaling protein.

Leaping forward a generation, the "hallmarks of cancer," as put forward by Robert Weinberg and Doug Hanahan in 2000, were derived from decades of observation and experimental studies, such as with *src*, and were widely accepted across the research and medical communities.[47] The two most prominent of these were the presence of activated oncogenes and inactivation or absence of tumor suppressor genes. There were signs of hope that drug discovery efforts aimed at stopping oncogenes by using "targeted therapies" would prevail over many cancer types. An important breakthrough in cancer treatment came from earlier research stemming from an oncogene discovered in the 1980s that resulted from a chromosomal translocation and known as *BCR-ABL*.[48] The gene fusion produced by the rearrangement causes chronic myeloid leukemia (CML), a rare blood cancer due to loss of growth control and cell death signaling (apoptosis)—two additional cancer hallmarks. The Swiss pharmaceutical giant Novartis had screened compounds for protein tyrosine inhibitors to target the ABL oncogene and discovered imatinib mesylate.[49] The small molecule was also found to be active against other protein tyrosine kinases (c-KIT and PDGFRα) and possibly able to slow tumors driven by mutations in those genes. In 2001, Novartis won FDA approval for the first precision drug to treat CML, which was given the brand name Gleevec in the United States.[50]

The success of Gleevec in turning CML into a chronic, rather than fatal, disease propelled research and drug discovery efforts to identify new classes of drugs with molecular specificity. As Gleevec was introduced to the market, the sequence of the first human genome became available in 2001 (the initial draft was published in 2001; the completed version in 2003) and provided drug hunters with a vast catalog of potentially new drug targets. However, the trove of human genome information on its own did little to aid in cancer research. Without an understanding of the functions of the genes identified through genomics, and with no further insights into processes that governed cancer growth, progress stalled. Investigators still focused on individual genes, and clinicians continued to view cancer through the lens of anatomy (for example, breast, lung, or liver cancer), not by shared, underlying molecular features.

The next stage in the progression toward a more quantitative and computational approach to cancer research, alas, came from the field of genomics. Massively parallel next-generation sequencing had matured by 2008, and Elaine Mardis and Richard Wilson at the Genome Center at Washington University in St. Louis had proposed in a National Institutes of Health (NIH) research grant to sequence entire cancer genomes. Rather than testing individual gene hypotheses for a given cancer type, comprehensive sequencing of a tumor would essentially provide an unbiased way to uncover molecular changes. This would be a hunt for somatic mutations—genetic changes occurring post-birth in cancer prone tissues. Since automated DNA sequencing technology and informatics tools were ready, and the data from normal human genome sequences had been less informative regarding cancer, Mardis and Wilson hoped that the study would prototype a new approach, referred to as *tumor/normal sequencing*. Grant reviewers felt otherwise, suggesting rather forcefully that a massive sequencing project on tumor DNA, potentially costing $1,000,000, would be less useful than careful study of individual genes, as had been the traditional approach of the past two decades.

Disregarding the rejection, Mardis, Wilson, and colleagues at the Genome Center forged ahead, sequencing DNA from a patient with acute myeloid leukemia on Illumina's latest Genome Analyzer instrument. The historic paper was published in the journal *Nature* in 2008. With astonishing accuracy, the NGS techniques had first identified a collection of 3,813,205 SNPs detected by comparison of the patient's tumor to the individual's healthy skin genome. Next, utilizing a computational analysis pipeline to eliminate those SNPs present as naturally occurring variants and not specific to the tumor, they uncovered (and independently validated) eight acquired somatic mutations. The final sentence of the abstract provided an exclamation mark to the shortsightedness of the grant reviewers: "Our study establishes whole-genome sequencing as an unbiased method for discovering cancer-initiating mutations in previously unidentified genes that may respond to targeted therapies.[51]"

Over the course of the next decade, thousands of cancer genomes were sequenced through efforts such as the Cancer Genome Atlas (TGCA)[6] and the International Cancer Genome Consortium (ICGC).[7] In parallel, a whole new industry around cancer genomics took shape to provide diagnostics based on established DNA mutations and molecular signatures or biomarkers from tumor-specific gene expression profiles and cell surface antigens. Incredible data resources have been developed, based on genomic surveys (whole genome and whole exome sequencing, plus targeted sequencing). These include ClinVar, dbGAP, and Catalog of Somatic Mutations In Cancer (COSMIC).[52,53] The COSMIC database began in 2004 as a scientific, literature-based collection of all published data on cancer samples and mutations. In its first year, the Wellcome Sanger Institute project had curated a set of 66,634 samples and 10,647 mutations. By 2018, the COSMIC database had grown enormously, adding 1.4 million samples along with 6 million mutations. By analyzing the massive collection, researchers have proposed that 223 key cancer genes drive nearly all 200 types of human cancers.[54]

Paradoxically, for most human cancers, treatments targeting these specific cancer genes or associated pathways are still unavailable. The pharmaceutical industry has been remarkably unsuccessful in discovering new therapeutics to treat cancer, with success rates hovering around 10 percent (also known as an *attrition rate*: the percentage failure at clinical trial stages is 90 percent) in most drug discovery programs worldwide.

The consensus among pharmaceutical executives is that several problems needed to be addressed in oncology R&D to increase chances for wins in clinical pipelines and for patients facing harsh regimens of chemotherapy and surgery in their battles against a myriad of cancers. Once immunotherapy drugs entered the scene with miraculous results in a few cancer types, an avalanche of interest and investment went toward this drug class and the newfound ability to genetically engineer T cells (CAR-T therapies; see Chapter 4). The classic small-molecule, target-based approach was in dire need of fresh ammunition in the form of functional genomics data. Why weren't these drugs working? The top concerns were to address whether the protein targets themselves (the oncogenic drivers discovered from tumors) were suitable and to determine what could be done to improve the clinical efficacy of anti-cancer agents. Since a majority of drug screening assays were performed in cancer cell lines, were these models understood at the molecular level? Comprehensive molecular profiling, similar to what was being done in primary tumors, was needed. It was also understood that certain mutational profiles might be more susceptible and thus more sensitive to drugs that were known to engage a target successfully, but that did not produce clinical results. This idea is individualized, precision medicine—getting "the right drug to the right patient at the right time" to achieve better results. From the collected genomic, epigenomic, and clinical data, it was hoped that further insight would help predict drug responses and redirect anti-cancer compounds to more tailored applications.

A leading scientific team in the United Kingdom took a data-driven scientific approach to integrate functional genomic profiling with drug screening and then use machine learning to predict which features of the cancer cell predict drug responsiveness.[55] The framework developed by Matthew Garnett's group highlights the process and quantitative methods used. Taking human cancer patient tumor genome data from 11,289 samples, algorithms identified thousands of cancer functional events of clinical importance. These were broadly classified into mutations, amplifications, and deletions, and gene promoter hypermethylation sites—an important form of epigenetic modification in cancer. This same multi-omic approach plus gene expression profiling (transcriptomics; see the "Omics Technologies and Systems Biology" section) was used to evaluate more than 1,000 cancer cell lines that have been derived from tumors. From these data, a status matrix could be built from the 1,000 cell lines and a condensed set of cancer functional events from the multi-omic profiling. Comparison of the primary tumor data to the cell lines revealed substantial capture of relevant cancer mutations across the cell line panel, setting the stage for drug sensitivity screening in these molecularly characterized in vitro models.

The large-scale pharmacogenomic profiling experiment was launched to measure drug effects on cell viability for a set of 265 compounds across every cell line, resulting in more than 1,000,000 data points generated from more than 200,000 dose response curves (~5 data points per compound). The data from each experiment (the derived IC50 value) was then fed into their quantitative framework employing statistics and machine learning techniques. The outputs allowed the researchers to make inferences about what class of drugs best target a particular cancer event and also which data types are most predictive of drug sensitivity. The pharmacological modeling revealed a large set of cancer-specific, pharmacogenomic interactions. The machine learning models indicated that genomic features (cancer driver mutations and gene amplifications) provided the best predictive models and could be improved with DNA methylation data, but not gene expression data, for specific cancer types. The pharmacological models had immediate clinical implications, identifying potentially new treatments that can be tested in clinical trials. The machine learning models suggested that clinical diagnostics should be focused on detecting underlying DNA alterations versus other molecular signatures of the tumor (namely, DNA methylation or gene expression).

A more powerful way to investigate individual gene contributions to cancer phenotypes and drug responses is to utilize CRISPR-Cas9 (see

Chapter 4) genome-scale screens.[56] This molecular genetic approach is another example of unbiased, hypothesis-free searches for insights. The technology can be used to activate, mutate, or silence (knockout) single genes with exquisite precision. While particular genes or pathways are examined in focused studies, a genome-scale screen interrogates every gene and potentially other functional elements in the genome. The utility of CRISPR systems in cell line models is particularly appealing, since loss of function screens can quickly uncover drug targets and cellular pathways critical in maintaining cancer cells' transformative properties. Algorithms can process the information and work to prioritize cancer drug targets.

Studies of this type are bringing computational cancer research in full view. Behan and colleagues designed genome-scale CRISPR-Cas9 viability screens to identify genes essential for cancer cell survival.[57] A set of 324 cancer cell lines were chosen, and 18,006 genes were targeted for knockout. The study ran more than 900 experiments in their pipeline and assigned fitness scores to genes based on the survival assays. In each cell line, a median of 1,459 fitness genes were found, an overwhelming number to explore manually. Within the fitness gene set, they used a computational technique, ADaM, to distinguish criteria for core fitness genes and those that are cancer-type specific. The method assigned core fitness genes as those present in 12 out of the 13 cancer types (for example, breast, CNS, pancreas) represented by the cell lines. The core set thus represents "pan-cancer" fitness genes, and many of these, as expected, were previously identified as essential genes (399/533), yielding a newly discovered set of 123 housekeeping genes involved in essential cellular processes. For the cancer-specific set, the computational process classified 866 as core fitness genes. From both of these sets, further analysis resulted in a prioritized set of 628 potentially new drug targets, with a majority (74 percent) specific to one or two cancer types, a remarkable outcome from the analysis pipeline. The scale and yield of this computationally driven research is indeed breathtaking; it has provided a target-poor area of drug development with both the tools for exploration and therapeutic hypotheses to test in relevant in vitro and in vivo cancer models.

Structural Biology and Genomics

Data science and computation are the driving force in structural biology. Genomics and structural biology work hand in hand, providing key data to allow drug developers to explore rationally the most promising ways to target critical protein components or alter complex cellular pathways.

Within hours of the SARS-CoV-2 genomic data being uploaded by Chinese scientists in Beijing, researchers around the world were able to use cloud computing to analyze the sequence, design experiments, and physically manufacture genes and proteins for further characterization.

When scientists talk about protein structure, they are usually referring to secondary or tertiary structure that is a consequence of how a protein folds in nature. All proteins are made up of a string of amino acids, taken from a universal set of 20, and each protein's unique 3D structure determines its biological function. Primary structure is simply the ordered sequence of amino acids. Secondary structure forms from patterns dictated by the primary sequence, usually repeated sequences that cause protein domains to form into helices and various sheet formations.

To get to 3D structural data, you first have to isolate the gene and produce a protein. Standard molecular biology protocols are used first to amplify and clone gene segments. Recombinant proteins are produced by inserting the cloned material into the bacteria (or in some cases yeast or other cells), and cultures are grown. The protein is purified, cryogenically frozen, and placed under an electron microscope for cryo-EM or crystallized and examined by X-ray crystallographic methods.

By the first few months of 2020, research groups had navigated from viral genome sequence to atomic-scale, 3D structures for three of the most important pharmaceutical drug and vaccine targets of SARS-CoV-2: the spike glycoprotein, the MPro protease, and RNA-dependent RNA polymerase. Although SARS-CoV-2 harbors one of the largest RNA virus genomes ever discovered, it encodes fewer than 30 proteins. In contrast, a prokaryotic genome such as *Escherichia coli* (*E. coli*, present in the human microbiome) possesses about 5,000 genes, and organisms such as flies, horses, and humans contain somewhere in the neighborhood of 15,000 to 30,000 protein coding genes.

With SARS-CoV-2, a spectacularly detailed image of the RNA-dependent RNA polymerase in complex with its associated proteins, nsp7 and nsp8, was derived after processing 7,994 micrograph movies using the cryo-EM technique.[58] At 2.9-Ångstrom resolution (the diameter of a water molecule is 2.75Å), the model provides a structural basis for how one inhibitor, remdesivir, incorporates into the complex and offers clues for the design of other candidate antiviral drugs. Similar insights were taken from structure of the spike protein in a trimer conformation, as shown in Figure 1.2, the viral surface protein essential for virions to bind to its cell surface receptor ACE2,[59] and the M protease, a proteolytic processing enzyme that is necessary for cleaving and releasing mature protein fragments from the longer virus polyprotein.[60] It is likely that additional, high-resolution structures of drug binding domains or antigen-antibody complexes will emerge in coming months to support R&D efforts to treat COVID-19.

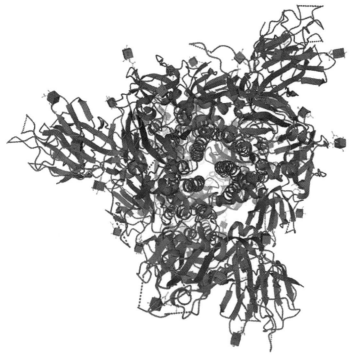

Figure 1.2: SARS-CoV-2 Spike glycoprotein structure *Source*: Image from Protein Data Bank[60]

Purely computational approaches to predict 3D protein structure from linear 1D sequence is the holy grail of structural biology. Prior to the availability of thousands of compute nodes on server farms to handle a protein folding algorithm's computational requirements, there were efforts such as Washington University's Folding@home, which started in Vijay Pande's laboratory at Stanford in 2000 and enlisted volunteer desktop CPUs for distributed jobs.[61] The team has since published hundreds of papers and has begun to churn out some impressive structures from the SARS-CoV-2 genome. A team at the company DeepMind (acquired by Google in 2015) has built AlphaFold and published the first use of deep learning on the protein structure prediction problem.[62] The most powerful aspect of DeepMind's work is that the algorithm works without relying on homologous templates to model preliminary structures. The core component of AlphaFold is a convolutional neural network that uses structures in a protein data bank (PDB) as the training set to predict the distances between carbon atoms of pairs of residues in a protein.

The "template-free" or *ab initio* free modeling is now yielding some partial structures of SARS-CoV-2 proteins, as published on DeepMind's website.[63] DeepMind and Folding@home are a small sampling of the large number of computational groups attempting a variety of novel methods to describe 3D structures accurately. Every year, the Critical Assessment of protein Structure Prediction (CASP) challenge has drawn 50 to 100 teams into modeling competitions. In the free modeling category, AlphaFold's performance at CASP13 surpassed expectations and appears to be bending the progress curve upward. As vaccine and therapeutic development take center stage in the pharmaceutical world, the molecular structures produced *in silico* may have enormous impact on the public health across the globe.

Sequencing the Human Genome

> *What biology does need, however, is a massive information base—a detailed knowledge of the genetic structure of several key organisms, including, for obvious reasons, man.*
>
> — Robert Sinsheimer, The Santa Cruz Workshop, May 1985

The Manhattan Project was launched after a single eminent scientist, Albert Einstein, informed Franklin Roosevelt in a now famous 1939 letter on the technological feasibility of producing powerful bombs based on his knowledge of science happening across Europe. The top-secret project led to the development of two alternatives to create the nuclear chain reaction needed, each of which was deployed in the atomic bombs dropped on Nagasaki and Hiroshima at the end of World War II. The Manhattan Project, with its organization and engineering focus, became the forerunner and model of all future "Big Science" projects initiated in the United States under leadership of the Department of Energy (DOE).

Following the birth of biotechnology and building on progress in human genetics, a proposal for sequencing the entire human genome was put forward and debated at workshops and conferences by prominent biologists and DOE administrators at the Office of Health and Environmental Research starting around 1985.[1,63] The project, dubbed the Human Genome Initiative by the DOE, was to provide a valuable resource for biology and medicine and demonstrate American competitiveness in science. As part of DOE's mandate to assess the health risks of atomic weapons and energy sources, a human genome reference sequence would be ideal for evaluating genetic damage caused by radiation and energy transmissions. It was an audacious goal for biological science, and unlike the Manhattan Project, no clear technology roadmap existed for any phase of the project.

The workshop attendees at the Santa Cruz meeting in 1985 sketched out several requirements, and at least three critical technology components were missing but necessary to tackle sequencing of the human genome. First, molecular techniques were needed to construct the physical and genetic maps of the genome. Genetic mapping techniques pioneered by David Botstein were a cause for optimism and were becoming an established molecular approach for locating disease-causing genes in human genomes.[64] Physical mapping efforts underway in two laboratories at the time, Maynard Olson's at Washington University working on the yeast genome and John Sulston's at Cambridge University on nematode, provided evidence of technical feasibility but were possibly too cumbersome for large genomes.[65,66] Olson had developed some of the first computational algorithms to decipher restriction enzyme digests of DNA to derive mapping information.

Automated DNA sequencing technology would be essential but did not yet exist outside of one laboratory. The first prototype DNA sequencing instrument had just been built in 1985 in Leroy Hood's group at the California Institute of Technology in Pasadena.[67] Related to the sequencing strategy, there was only the vaguest of ideas about what computational approaches would be needed to aid in the sequencing and reconstruct the genome sequence, or how to process, store, and analyze the information. The first nucleotide sequence matching and alignment algorithms were terribly slow and would not scale well for high-throughput sequence analysis. The most important algorithms, those to evaluate sequence quality and assemble the pieces correctly into a genome, were many years away from development. More importantly, it wasn't even clear what sequencing strategies were going to be used or the nature of the computational problems that such a project would need to solve. The largest genome sequenced by 1985 had been that of the Epstein-Barr virus, at 1.72×10^5 bases. The human genome was estimated at 3×10^9 bases. To use a travel analogy, if the goal was to reach the sun at a distance of 90 million miles away from Earth, the project had made it 5,500 miles from San Francisco to Paris by commercial aviation and now had to build a spacecraft for the rest of the journey. The Big Science project of biology seemed like a pipe dream to the skeptics.

At the Santa Cruz meeting held in 1985, the 12 experts were split evenly: half were in favor, and half were opposed to the project. Botstein, the Massachusetts Institute of Technology geneticist, roundly condemned the project. Scientists, including Botstein, were worried that Big Science funding would draw money away from small investigator RO1 awards typically received from the NIH. Other critics of the project, some in the scientific community and others in Congress, objected to projected costs, anywhere from $1 billion to $3 billion dollars over 15 years. There was unease at the politicization of science, how the effort would be organized, and who would set the scientific priorities. A few researchers saw little benefit in deriving the chromosomal DNA sequence because they were mainly interested in the protein coding sequence contained in exons, which would not be easily inferred from genomic DNA due to the intron/exon structure of eukaryotic genes. Coding sequence could be obtained only from analysis of mRNA transcripts, as gene finding algorithms were not an anticipated development out of the genome project.

The molecular biologists and human geneticists were generally enthusiastic about the project and had grandiose visions of medical breakthroughs and acquiring a deeper understanding of human nature, all emanating from a full reading of our species' genomic sequence. The reductionist mindset that was prevalent at the time viewed the detailed description of every gene that would be provided by the genome sequence to be the key that unlocked all of biology's secrets. The geneticist's view of medicine was also seen through the lens of the genome and predicted that the underlying cause of all of the 3,000 or so known genetic disorders would be solved through sequencing and a catalog of genetic variation. These views turned out to be overly simplistic. For example, even with extensive family and medical histories combined with known genetic changes found in individuals with the psychiatric disorder schizophrenia, the cellular basis of the neuropathology remains poorly understood, and therapeutic approaches targeting various neurotransmitter systems have failed.

Among the enthusiasts for a genome initiative was Walter Gilbert, who had just returned to Harvard University after running Biogen, one of the first biotechnology companies formed to develop therapeutics based on recombinant DNA technology. He was unconvinced that a government effort would ever get off the ground and had been a provocative member of the National Academy of Sciences committee formed by James Watson to explore a human genome project.[68] Gilbert stepped down from the committee and began looking to privatize an effort to sequence the genome by seeking venture capital for his new brainchild, a startup called the Genome Corporation.[69] It was a controversial move for several reasons, and his first attempt at running a biotechnology company had ended poorly in 1984, as he was forced to resign as CEO from Biogen. The company had lost money in the two previous years, it was competing on the same biotech turf with savvy Genentech in San Francisco, and nervous investors were looking for seasoned business leadership. Apart from his business endeavors, Gilbert was a brilliant, innovative scientist who had made fundamental discoveries in molecular biology since the early 1960s; he had won a share of the Nobel Prize in Chemistry with Frederick Sanger in 1980 for developing DNA sequencing technology.

Gilbert's zeal to press a human genome sequencing initiative of his own forward and turn gene sequence information into business profits was alarming to many and raised ethical issues. How could a private enterprise own the genome's information? If you owned a newly sequenced fragment of DNA, was that a novel "composition of matter" and sufficient for a patent? For venture capitalists, they couldn't readily see a business model that would generate revenue and the size of the gene data market was unknown. Ultimately, the Genome Corporation never attained liftoff, due in large part to the stock market crash in October 1987. However, a decade later, the genome community was shocked to learn that in the midst of its sequencing efforts, biologist and biotech entrepreneur Craig Venter had formed a private company, Celera Corporation, and would set out to race the publicly funded effort to the finish line, base by base, to commercialize the genome.

Although there was less enthusiasm in the larger scientific community supported by the NIH, Congress was clearly behind the moonshot and appropriated funds for the Human Genome Project in 1988. The money went to both the DOE and NIH, whom together had agreed to jointly manage the effort. The draft plan was published in April 1990 with a five-year aim to develop technology, followed by a second phase projected to produce a completed genome in 2005. The estimated budget was $3 billion over the project's lifetime. The work of the DOE would be initiated by three national laboratories with excellent track records for technology development. These were Lawrence Livermore, Lawrence

Berkeley, and Los Alamos National Laboratories. The NIH had set up a separate entity to oversee the project, the Office of Human Genome Research, which was run part-time by Watson. Another strategic move was to include laboratories working on genome sequencing at top centers worldwide. The plan was to work together as the International Human Genome Sequencing Consortium. Over time, the NIH established the National Center for Human Genome Research, headed by Francis Collins after Watson's departure in 1992. The Human Genome Project was formally completed in 2003 with an estimated price tag of $2.7 billion—two years early and $300 million under budget.

The monumental achievements of the Human Genome Project are now etched in history and well-documented since the landmark publications of the draft sequence in 2001.[70,71] The success of the Herculean efforts in both public and private spheres originated from its conceptualization as an engineering project from the start. The DOE's raison d'être is management of big-budget technology development megaprojects, from telescopes to high-energy physics installations. Over the timespan of the project, a steady stream of technology innovation proceeded across DNA sequencing instrumentation, DNA sequence assembly strategies, and project data coordination among the genome centers. The impetus to sequence the three billion base pair human genome drove exponential increases in sequencing throughput, lowered costs, and led to completion of many important genomics projects in parallel.

What was not appreciated at the onset of the Human Genome Project was that the initial focus on sequencing of overlapping, contiguous DNA fragments contained in an array of clones was never going to be speedy enough to finish the genome in a decade. Only with the breakthrough of whole genome shotgun sequencing strategies, a concept initially suggested as early as 1981 and later developed at scale by Craig Venter at the Institute for Genome Research (TIGR),[72,73] would the genome get completed. TIGR's groundbreaking announcement in 1995 of the sequencing of the bacterium *H. influenza* with shotgun methods illustrates the take-off in sequencing power, as shown in Table 1.2. The progress over time from when the first DNA molecule was sequenced in 1965 to beyond the sequencing of the first human genome was driven first by chemistry, then by instrumentation improvements, and finally by massively parallel sequencing with high-performance computing to match.

Table 1.2: Milestones in DNA Sequencing—From Single Genes to Metagenomes

ACHIEVEMENT	MOLECULE OR GENOME	YEAR	SIZE (BASES)
First DNA molecule	Alanine tRNA	1965	77
First gene	Phage MS2 coat protein	1972	417
Bacteriophage RNA	RNA sequence MS2	1976	3,569
Bacteriophage DNA	DNA sequence phiX174	1977	5,386
Viral genome	SV40	1978	5,200
Bacteriophage	Lambda	1977	48,502
Organelle genome	Human mitochondria	1981	16,589
Viral genome	Epstein-Barr virus	1984	172,282
Bacterial genome	*Haemophilus influenza*	1995	1,830,137
Yeast genome	*Saccharomyces cerevisiae*	1996	12,156,677
First multicellular organism	Nematode C. elegans	1998	100,000,000
Genetic model invertebrate	Fruit fly D. melanogaster	2000	120,000,000
Genetic model plant	Mustard weed A. thalania	2000	135,000,000
Vertebrate genome	*Homo sapiens*	2001	3,000,000,000
genetic model vertebrate	Mouse *mus musculus*	2002	2,500,000,000
Neanderthal genome (ancient DNA)	*Homo neanderthalensis*	2010	3,000,000,000
Metagenome	Human microbiome	2012	22,800,000,000

Bioinformatics and computational biology techniques were developed *de novo* throughout the lifetime of the Human Genome Project. The importance of information technology was on display as the genome neared completion, with assembly tasks requiring heavy CPU demands and RAM memory. The lasting benefit of the project was the establishment of a new infrastructure—a new framework for biological sciences. Scientists could now begin to look at problems with an unbiased genome-wide view, not necessarily locked into testing hypotheses on subset of genes, variants, or selected cellular mechanisms decided in advance, prior to experimentation. The human reference sequence was the prerequisite to unlocking the gates to discoveries spanning across fields from anthropology to zoology.

In retrospect, the powerful capabilities of the information revolution were the enabling gifts of the high-tech sector to biology. Several of the chief scientists directing the project reminisced in 2003 about the importance of computation. David Botstein said the most surprising aspect of the project's completion was that without computers, there would not have been a Human Genome Project. Maynard Olson said that "the whole computational infrastructure didn't exist."[74] For David Haussler at the University of California, Santa Cruz, it was the importance of the work of his colleague Jim Kent, who was responsible for the final genome assembly and also the key developer of the software for both the assembler and genome browser.[75,76] The grandeur of the Human Genome Project left Haussler waxing poetic:

> And we realized—we had this feeling of walking into history, this is it! This is the world—the whole world is seeing its genetic heritage for the first time. Humankind is a product of 3.8 billion years of evolution. Here is this amazing sequence of information that had been crafted by all of the great triumphs and stumbles of all of our ancestors over so many eons and there it was. And for the first time we're reading it, we can actually read the script of life that was passed down to us.[74]

Computational Biology in the Twenty-First Century

The staggering complexity of the human genome did not come into full view upon its completion in 2003. From today's vantage point, the concept of the genome as merely an information storage site housed in a nuclear neighborhood containing 23 chromosomal pairs is misleading.

This antiquated picture implied that the instructions to guide cellular processes are retrieved from a static genome, much in the same way that a computer performs operations by loading software instructions into memory and running hardware. The genome as software is certainly by now an outdated metaphor.

The static view of the genome stemmed from a lack of understanding of how events, such as gene transcription, DNA replication and repair, and epigenetic events, are regulated. Molecular studies in the post-genome era have revealed an interactive array of regulatory molecules and processes directing a highly dynamic, 3D genome. New regulatory mechanisms and classes of noncoding RNAs were discovered, which include micro RNAs (miRNAs), small interfering RNAs (siRNAs), and long, noncoding RNAs (lncRNAs) that are transcribed and function mainly to inhibit gene expression. A gene silencing pathway, known as *RNA interference (RNAi)*, is triggered by the presence of siRNAs or miRNAs to repress gene expression at multiple levels. DNA replication and transcriptional regulation is tightly governed by access control to chromosomal regions in either "closed" heterochromatin or "open" euchromatin states. RNAi is important in replication and represses expression from transposable elements that occupy half of the genome.

Another important feature of genome structure and regulation, known for decades, is the interaction of very distal enhancer elements with proximal gene promoters that are often found 100 kilobases away from each other. To impact gene activity, these two genomic features must move in close proximity within 3D genome space. There, the gene-activating transcription factors on the enhancer can switch on the promoter near the gene's transcription start site. More recently, several facets of the 3D genome have come to light. A significant fraction of the genome (30–40 percent) interacts with the nuclear lamina, where heterochromatin regions containing as much as 10 megabases of DNA form lamina-associated domains (LADs).[77] Physical interaction in the LADs appears to prevent access to gene promoters, uncovering yet another transcriptional repression mechanism.

Finally, new technologies are revealing the nature and extent of epigenetic modifications across the genome. Environmental influences or experiences can cause heritable changes to the genome via epigenetic mechanisms. They work by modifying DNA or histone proteins bound to DNA and alter gene activity instead of gene sequence. Epigenetic changes are achieved directly by methylation of cytosine bases (DNA methylation) or by histone modification such as reversible acetylation and methylation to modulate genome function. A new picture thus emerges of a dynamic, 3D genome interacting with thousands of molecules, entering and exiting nuclear subdomains, to control expression of thousands of distinct gene repertoires whose protein components carry out all of the cell's biochemical processes. Francis Crick's powerful framework of molecular biology—the central dogma stating that DNA makes RNA makes protein—has the added dimension of feedback regulation. Thus, a more appropriate metaphor used in the past decade considered the genome as an RNA machine.[78] But the genome stores memories from epigenetic modifications and transmits data to the next generation in cycles of recombination and replication. A more encompassing view is that the genome is a molecular machine for information processing.

Applications of Human Genome Sequencing

For most computational biologists, the human reference genome is an afterthought—a relatively small, 3.275 gigabyte (GB) file that can be readily downloaded from the Internet in a minute. Your child's high school biology class probably includes a day of surfing a website and taking a look at genes and genetic variation in a genome browser, likely built by the Broad Institute,[79,80] the UCSC Genomics Institute,[76,81] or the Wellcome Sanger Institute[82,83] in the United Kingdom. Researchers pore over the genome to evaluate the rich annotation now available, where at any position of genome, information can be obtained on location of functional elements like exons or introns, sites of epigenetic modification, transcript identity, and known variants of clinical importance.

It's amazing to think that $3 billion worth of production sequencing and technology development is freely available for anyone to utilize and explore. This happens across industries, as the end result of complex engineering feats for commercial products, such as an iPhone, newly developed drugs, and the graphics processor for your desktop computer, probably cost each individual corporation well over a billion dollars to develop. The products are built for massive consumer markets, and the costs are amortized over many years with comfortable profit margins, making the long-term investments worthwhile. For taxpayer-funded megaprojects, research communities worldwide reap the benefits in the hope of transferring the scientific knowledge back to society in a beneficial way.

Over the course of the Human Genome Project, the many genome centers and research laboratories that built DNA sequencing factories produced many valuable genomes (refer to Table 1.2 again for some of their major achievements). Far beyond the initial sequencing of model organisms, the research community has also completed thousands of eukaryotic genomes and tens of thousands of bacterial genomes. Enormous amounts of experimental sequencing data from omics technologies (see the next section, "Analyzing Human Genome Sequence Information") has likewise been deposited into nucleotide sequence repositories, such as those found at the National Center for Biotechnology Information, or NCBI.

Producing finished genome sequences of new species is still somewhat of an art, requiring computational skill for genome assembly and sequence annotation. However, DNA sequencing of individual human genomes is now done routinely; the data that is captured is not an assembly, but a base-by-base comparison to updated versions of the human reference.

DNA SEQUENCING APPLICATIONS FOCUS ON GENETIC VARIATION IN HEALTH AND DISEASE

These applications utilize several modes of detecting variants in genome sequence: whole genome sequencing (WGS) involves analysis of the entire genome; whole exome sequencing (WES) captures protein-coding exons across the genome; NGS-based liquid biopsy analyzes DNA found in biological fluids; and targeted sequencing examines only selected genes in defined subsets or diagnostic panels. The following is a list of the variety of ways that human genome sequencing is being used in the clinic, by consumers, and by researchers and the pharmaceutical industry:

Rare disease diagnostic sequencing: WGS used to pinpoint inherited disease mutations, often together with parental sequences

Cancer diagnostic sequencing: WGS, WES, or targeted approaches used in clinical evaluations of tumors as compared to normal genome; liquid biopsy and sequencing of blood, urine samples

Prenatal diagnostic sequencing: Use of noninvasive prenatal testing to detect chromosomal anomalies (primarily trisomy 21)

Personal genome sequencing: WGS or SNP arrays for direct-to-consumer service for identifying traits, health risks, and research participation

Population studies: WGS and WES for basic research to aid in disease gene discovery, population allele frequencies, and human evolution and migration patterns

Case-control studies: WGS and WES for basic and clinical research to identify variants associated with traits or disease

Population health programs: WGS in countries for public health combined with medical records

Pharmacogenomic studies: WGS, WES, or targeted sequencing to evaluate genotype:drug associations for drug development

Patient stratification: WGS or WES to sequence individuals prior to clinical trials for grouping by genotype

Pharmaceutical research: WGS and WES for drug target identification and target validation by identifying risk alleles, genetic modifiers, and loss of function variants underlying disease mechanisms

Genetic variation is also detected with technologies that capture "genotypes" at defined sites using SNP chip or bead arrays.

Analyzing Human Genome Sequence Information

The myriad applications of human genome sequencing drive toward the twin goals of identifying the genetic basis of disease and cataloging variation in individuals and populations. The tools to accomplish this come out of computational biology, which are rooted in statistics, mathematics, and computer science. When the Human Genome Project was completed, and for several years thereafter, few envisioned that genome sequencing would become routine with a production cost of around $1,000.

As was discussed with the sequencing of SARS-CoV-2, NGS technology was the breakthrough entryway. The NGS method commercialized by Illumina combines engineering of ultra-high density nanowell arrays with sequencing by synthesis chemistry, where billions of reactions can occur in parallel. The technology utilizes the shotgun sequencing strategy, where DNA is sheared into tens of millions of fragments and a "library" is constructed. The instrument sequences fragments typically of 100 to 300bp in length, producing billions of sequence "reads" that are later aligned to the human reference sequence. The power of the method is that the entire genome can theoretically be covered at a level of redundancy that enables use of statistical techniques to determine accurately the sequence and resulting genotypes derived from the two copies present in a diploid genome. For WGS and WES, a genome coverage level of 30X to 50X is required. Figure 1.3 shows a diagram of a modern analysis pipeline for processing NGS data.

Processing NGS data proceeds from primary analysis on the instrument (for example, generating base calls) to secondary analysis that maps and aligns reads to the reference, followed by a multistage variant calling process, and finally a tertiary annotation and interpretation step.

In NGS data processing, the read alignment step is usually the most computationally intensive. Here is a walk-through example of the difficulty: Suppose that there are 1 billion reads that need to be mapped back to their originating positions on the genome. Conceptually, each read could be evaluated at every position along the genome in stepwise fashion. For one 100bp read, that would be 3×10^9 comparisons, moving a single base at a time. Such a poor search strategy would result in 3×10^{18} operations and be highly inefficient. An exact match may not be found; what about those with 1, 2, or 3 mismatches per 100bp? Where do you store that staggering amount of information even temporarily? The way around this is to build search indexes, analogous to constructing a phone book such that partial matching sequences can be looked up and top hits stored for each read. Search algorithms operate this way, utilizing indexes and filtering steps to avoid computational explosions.[84] These search algorithms are usually built in combination with alignment programs that can tolerate gaps and mismatches and produce mapping scores and other metrics that are stored with the alignment information in enormous binary alignment or "bam" files.

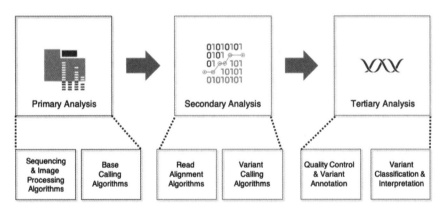

Figure 1.3: Genome analysis pipeline for whole genome sequencing on Illumina instrumentation

Capturing sequence and structural differences in an individual genome, recording the genotypes present, and assessing the quality of the data are the central tasks at the variant calling stage. Computational biologists have spent more than a decade improving the suite of tools necessary to handle the complexity of the tasks and the variables surrounding NGS data. Algorithms must determine zygosity—whether a genotype is present in either a homozygous or heterozygous state—and are tuned to handle specific variant classes, including the following:

- Germline and somatic single nucleotide variants (SNVs)
- Insertions and deletions (INDELS)
- Copy number variations (CNVs)
- Structural variants (SVs)
- Chromosomal breakpoint locations
- Gene fusions
- *De novo* mutations

Several of these tools utilize Bayesian statistical frameworks and leverage machine learning models to improve the accuracy of predicted genotypes.[85,86]

The fundamental issue that confronts researchers when analyzing these big datasets is how to discriminate a true variant residing in the genome from sequencing or mapping errors. Sequencing machine error rates hover around 0.25 percent.[87] Frequencies for genome variants are on par with base error rates or much lower: SNVs occur once every 500 to 1,000 bases (that is, 0.1 to 0.2 percent); CNVs and SVs are much rarer. Right away, you can see that sequencing errors are far more prevalent than expected occurrence of variants across the genome. Bayes' theorem has proven extremely powerful in predicting sites of variation. The simple mathematical formula in working with hypotheses (H) and data (D) for determining probabilities (P) in Bayes' theorem is as follows:

$$P(H|D) = P(H)P(D|H)/P(D)$$

When prospecting for variants at any site, there are four homozygous hypotheses to be considered: AA, CC, GG, TT, and 16 heterozygous combinations (AA, AT, AC, and so on). Genotype frequencies at a given site can be estimated from population databases, such as the 1,000 Genomes Project, which can be used to update the Bayes' prior probabilities in addition to the experimental sequence data from an individual sample. In variant callers, such as GATK's haplotype caller, genotype likelihoods, rather than probabilities, are calculated to estimate whether a hypothesis is true or false. Bayesian methods are employed widely in the genomics field, much in the same way that Bayesian learning and Bayesian networks have been used so successfully across many industries today.

Although the focus is often on personal genomes, variant discovery in research settings is done primarily in families and large populations. Discovery of new variants and detecting associations of genetic variation and disease often requires 10,000 or even 100,000 individual genomes! The reasons favoring large-scale genomics are that most common variants are already known (and nonpathogenic) and have been entered in databases, small cohort sizes in studies are prone to biases (for example, low ethnic diversity), and *de novo* mutations are hard to identify without family genomes to aid in discovery. New tools were developed for these types of studies, including "joint" variant calling algorithms that can leverage family information to derive genotypes more accurately and comprehensively. For joint calling algorithms, the Bayesian techniques perform the calculations simultaneously across all individuals. With the aid of the compute infrastructures available today, entire families and populations can be analyzed together.[88]

Omics Technologies and Systems Biology

The information revolution enabled the development of high-throughput DNA sequencing technology, helping to build the fields of genomics and computational biology. These disciplines either have directly created or have had a ripple effect on the advancement of many of the new "omics" technologies that are now in use across biology and drug development. Collectively, they acquire and deliver functional genomics data on different classes of molecules or molecular features in large-scale, discovery-oriented studies. As such, omics techniques have greatly expanded the shift toward unbiased approaches and quantitative methods in biology. In parallel, these omics datasets are being used by systems biology researchers toward the goal of achieving a more holistic view of physiology by integrating clinical, molecular, and cellular data along with population genetic information.

The conceptual framework of omics comprises a workflow that enables comprehensive molecular profiling of a distinct biological entity in a cell or tissue system. For each molecular type under investigation, a specific, stepwise process is followed from laboratory sample preparation, automated detection, and raw data collection on analytical instrumentation, to computational pipelines that process primary data and then generate insights by machine learning and statistical methods on one or more datasets. The genomics computational workflow was shown in the schematic in Figure 1.3. Standardized techniques for DNA and

RNA isolation exist for genomics and a variety of omics techniques discussed in a moment. Sample preparation for proteins is more varied and depends on the experimental goals. Standard operating procedures for sequencing instruments have been implemented for each of the various types of genomics applications. The community is far less standardized and much more active in developing new algorithms, statistical techniques, and visualization methods for the omics datasets.

The preceding sections of this chapter focused on the tight connection between the information revolution and the instrumentation and software developed to analyze the genome. No less important in biology, and with orders of magnitude greater complexity, is the proteome. Characterization of the proteome is necessary to gain insights into complex biological processes and phenotypes largely driven by protein functions; this information is not accessible with genomics data on its own.[89] Proteomics is challenging due to chemical and molecular characteristics not easily captured with a single technology. Proteins can acquire post-translational modifications, which include phosphorylation, acetylation, ubiquitinoylation, and lipidation. These modifications, with the exception of phosphorylation, are difficult to detect or quantify but are critical for determining functional states. The proteome is also hierarchically organized to carry out cellular functions. Sensitive techniques have been developed to detect a wide range of protein-protein interactions. Higher-order molecular complexes and interacting networks are not amenable to profiling *per se*, although they can be inferred from the data.

Mass spectrometry (MS) has been the mainstay analytical approach to identify any component of the proteome. Gaining an unbiased and complete coverage of the proteome is done using shotgun proteomics with data-dependent acquisition techniques, referred to as *DDA*. This is one of the experimental approaches of so-called bottom-up proteomics, where all peptides within a certain mass range are fragmented in tandem MS instruments, typically a quadrupole-orbitrap analyzer. In bottom-up studies, the protein sample is enzymatically processed prior to loading onto the spectrometer. The computational workflow in proteomics for DDA utilizes peptide identification algorithms that have undergone development for several decades. The most commonly used strategy is to search acquired tandem mass (MS/MS) spectra against a protein sequence database using database search algorithms. However, in spite of improvements in the quality of MS/MS data acquired on modern mass spectrometers, a large fraction of spectra remains unexplained.[90] The two other bottom-up methods are known as *targeted* and *data-independent acquisition* (DIA). In top-down proteomics, proteins are studied as intact entities, unprocessed. This has the advantage that all modifications that occur on the same molecule can (at least in principle) be measured together, enabling identification of the full-featured protein. The proteomic technologies utilizing MS are rapidly maturing and are in much more widespread use across the life sciences today than a decade ago. The completion of a proteome map will greatly enable targeted and DIA-based strategies because both rely on spectral libraries of known fragments for analysis. Further advances are taking shape in protein sequencing, single-cell proteomics, and novel informatics approaches to integrate proteomic and other omics data to decipher biological networks.

Of all of the more recent omics technologies, transcriptomics has arguably made the biggest impact in biological studies and drug discovery. Gene expression profiling technologies based on systematic PCR,[91] or microarrays,[92,93] were launched in the 1990s. A decade later, the ability to count RNA transcripts on NGS instruments led to a breakthrough application, RNA-seq.[94] A whole slew of new informatics tools was needed to deal with the complexities of RNA alternative splicing in order to count unique transcripts and for normalization of the data. The use of gene expression as a proxy for gene activity has been successful in studying many biological systems, and it was the first functional genomics method employing clustering algorithms and machine learning to evaluate patterns in the data. Genes co-expressed and coordinately regulated under a given cellular condition suggest that they function together. Identifying correlated patterns of activity and determining differential gene expression between normal and disease states has turned up important discoveries; in cancer studies, gene expression profiling routinely pinpoints regulatory pathways gone awry prior to any knowledge of underlying genetic mutations in the tumor sample. Genetic studies have also leveraged expression data to find genes and SNPs located within quantitative trait loci, termed eQTLs.

RNA measurements in traditional microarray or RNA-seq experiments use purified mRNA from bulk cells or tissues. In recent years, single-cell technologies have dramatically shifted momentum away from the averaged measurement resulting from bulk processing. The new way to describe biological activity is at single-cell resolution, which in turn has unlocked a whole new series of technologies in concert with single-cell RNAseq, or scRNA-seq.[95] Single-cell qPCR, mass cytometry, and flow cytometry enable the study of heterogeneity of cell populations. Single-cell transcriptomics has also gotten a boost from rapid innovation in visualizing transcripts *in situ*, usually on frozen or formaldehyde-fixed paraffin embedded (FFPE) tissue slides. Traditional nucleic acid techniques combined with fluorescent probes were limited to detecting one to four transcripts per slide. Newer spatial transcriptomics methods can visualize hundreds or even thousands of RNA species simultaneously with the use of oligonucleotide-based barcoding technology. The field is also expanding the repertoire of RNAs that can be detected, with platforms capable of measuring ribosomal RNAs, full-length RNAs (using long-read cDNA), and direct RNA counting or sequencing platforms.

Epigenomics arose almost in parallel with transcriptomics. The expression of genes is controlled at multiple levels with diverse mechanisms in eukaryotic cells. As discussed briefly in earlier sections, DNA regulatory elements are contacted by protein transcription factors, which bind at enhancers, promoters, and insulators to influence transcription of cognate genes. The ability to capture and analyze protein-DNA interactions, including the underlying DNA sequence, was enabled with CHIP-Seq. Epigenetic modifications can be profiled with several molecular techniques: Methyl-Seq for sites of DNA methylation and ATAC-seq for chromatin accessibility. Between transcriptomics and epigenomics, scientists have unparalleled data acquisition technologies to investigate the gene regulatory landscape.

Systems biology, and the fruits of the information revolution, are releasing a third wave of innovation: the tools for multi-omic data integration and analysis. Overlaying RNA-seq information with epigenetic data can reveal gene circuitry that may not be apparent on examination of datasets independently. Study of personal multi-omics in healthcare might provide unique, actionable insights, as was seen in pilot studies done at the Institute of Systems Biology. Systematically studying inter-omic, cross-sectional correlations is now possible.[96]

The analytical tools from computer science that are most promising for large-scale dataset evaluation reside in machine learning. In genomics, researchers at Google implemented deep learning to improve predictive performance of variant calling by using advances in computer vision and convolutional neural networks.[97] Machine learning has shown fundamental value for detecting underlying structure in omics datasets and is one of the few ways to achieve dimensionality reduction effectively, especially in scRNA-seq data.

For other omics types, the use of neural networks and deep learning may be limited due to the volume of data that can obtained for training the models. This need for data is compounded in genetic association studies, where identifying rare variants at statistically significant levels necessitates genotyping of millions, if not tens of millions of individuals. New data sharing paradigms that use cryptography could greatly aid in studies requiring population-scale datasets.[98] A similar problem arises in cancer prediction from omics data, but the issue is one of heterogeneity. Modern machine learning approaches will attempt to find recurrent patterns. But what if a similar outcome is found from heterogeneous inputs? With cancer, different genetic alterations lead to the same biological phenomena, a transformed cell. Many of these problems will need to be sorted out experimentally, rather than algorithmically.

Notes

1. National Research Council. Mapping and sequencing the human genome. Washington, D.C.: National Academy Press; 1988.

2. Legrain P, Aebersold R, Archakov A, Bairoch A, Bala K, Beretta L, et al. The human proteome project: current state and future direction. Mol Cell Proteomics MCP. 2011 Jul;10(7):M111.009993.

3. Turnbaugh PJ, Ley RE, Hamady M, Fraser-Liggett CM, Knight R, Gordon JI. The human microbiome project. Nature. 2007 Oct 18;449(7164):804–10.

4. Amunts K, Ebell C, Muller J, Telefont M, Knoll A, Lippert T. The Human Brain Project: Creating a European Research Infrastructure to Decode the Human Brain. Neuron. 2016 Nov 2;92(3):574–81.

5. Insel TR, Landis SC, Collins FS. Research priorities. The NIH BRAIN Initiative. Science. 2013 May 10;340(6133):687–8.

6. Cancer Genome Atlas Research Network, Weinstein JN, Collisson EA, Mills GB, Shaw KRM, Ozenberger BA, et al. The Cancer Genome Atlas Pan-Cancer analysis project. Nat Genet. 2013 Oct;45 (10):1113–20.

7. International Cancer Genome Consortium, Hudson TJ, Anderson W, Artez A, Barker AD, Bell C, et al. International network of cancer genome projects. Nature. 2010 Apr 15;464(7291):993–8.

8. All of Us Research Program Investigators, Denny JC, Rutter JL, Goldstein DB, Philippakis A, Smoller JW, et al. The "All of Us" Research Program. N Engl J Med. 2019 15;381(7):668–76.

9. Bycroft C, Freeman C, Petkova D, Band G, Elliott LT, Sharp K, et al. The UK Biobank resource with deep phenotyping and genomic data. Nature. 2018;562(7726):203–9.

10. GenomeAsia100K Consortium. The GenomeAsia 100K Project enables genetic discoveries across Asia. Nature. 2019;576(7785): 106–11.

11. Schwab K. The fourth industrial revolution. New York: Crown Business; 2017.

12. Meyer SL. China created a fail-safe system to track contagions. It failed. New York Times. 2020 Mar 29;

13. Yu G. In depth: How early signs of a SARS like virus were spotted spread and throttled in China. Caxin Global. 2020 Feb 29;

14. The Associated Press. China delayed releasing coronavirus info, frustrating WHO. 2020 Jun 2;

15. Page J, Fang W, Khan N. How it all started: China's early coronavirus missteps. Wall Street Journal. 2020 Mar 6;

16. Randolph J. Translation: Li Wenliang's "admonishment notice." China Digital Times. 2020 Feb 6;

17. Cho A. AI systems aim to sniff out coronavirus outbreaks. Science. 2020 22;368(6493):810–1.

18. Etzioni O, Decario N. AI can help scientists find a Covid-19 vaccine. Wired. 2020 Mar 28;

19. Wuhan Institute of Virology, website report [Internet]. 2020 Jan. Available from: `http://www.whiov.ac.cn/xwdt_105286/zhxw/ 202001/t20200129_5494574.html`

20. Zhou P, Yang X-L, Wang X-G, Hu B, Zhang L, Zhang W, et al. A pneumonia outbreak associated with a new coronavirus of probable bat origin. Nature. 2020 Mar;579(7798):270–3.

21. Wu F, Zhao S, Yu B, Chen Y-M, Wang W, Song Z-G, et al. A new coronavirus associated with human respiratory disease in China. Nature. 2020 Mar;579(7798):265–9.

22. Tan W, Zhao X, Ma X, Wang W, Niu P, Xu W, et al. A Novel Coronavirus Genome Identified in a Cluster of Pneumonia Cases — Wuhan, China 2019–2020. China CDC Wkly. 2020;2:61–2.

23. Holmes EC, Dudas G, Rambaut A, Andersen KG. The evolution of Ebola virus: Insights from the 2013–2016 epidemic. Nature. 2016 Oct;538(7624):193–200.

24. Gisaid.org [Internet]. Available from: https://www.gisaid.org/

25. Nextstrain.org [Internet]. Available from: https://nextstrain.org/

26. Bedford T. Cryptic transmission of novel coronavirus revealed by genomic epidemiology [Internet]. 2020. Available from: https://bedford.io/blog/ncov-cryptic-transmission/

27. Gallo RC, Montagnier L. The Discovery of HIV as the Cause of AIDS. N Engl J Med. 2003 Dec 11;349(24):2283–5.

28. Annual summary 1979: reported morbidity and mortality in the United States. Center for Disease Control; 1980. Report No.: 28(54).

29. TABLE I. Annual reported cases of notifiable diseases and rates per 100,000, excluding U.S. Territories- United States 2018. Centers for Disease Control; 2018.

30. Ksiazek TG, Erdman D, Goldsmith CS, Zaki SR, Peret T, Emery S, et al. A Novel Coronavirus Associated with Severe Acute Respiratory Syndrome. N Engl J Med. 2003 May 15;348(20):1953–66.

31. Assiri A, McGeer A, Perl TM, Price CS, Al Rabeeah AA, Cummings DAT, et al. Hospital outbreak of Middle East respiratory syndrome coronavirus. N Engl J Med. 2013 Aug 1;369(5):407–16.

32. Stadler T, Kühnert D, Rasmussen DA, du Plessis L. Insights into the early epidemic spread of ebola in Sierra Leone provided by viral sequence data. PLoS Curr. 2014 Oct 6;6.

33. Kleiber M. Body Size and Metabolism. Hilgardia. 1932;6(11):315–51.

34. White CR, Blackburn TM, Seymour RS. Phylogenetically informed analysis of the allometry of mammalian basal metabolic rate supports neither geometric nor quarter-power scaling. Evol Int J Org Evol. 2009 Oct;63(10):2658–67.

35. West GB, Brown JH, Enquist BJ. A general model for the origin of allometric scaling laws in biology. Science. 1997 Apr 4;276(5309):122–6.

36. West G. Scale: The universal laws of growth, innovation, sustainability, and the pace of life in organisms, cities, economies and companies. New York: Penguin Press; 2017.

37. Hawking S. Unified field theory is getting closer, Hawking predicts. San Jose Mercury News. 2000 Jan 23;

38. Illumina instrument [Internet]. Available from: https://www.illumina.com/content/dam/illumina-marketing/documents/products/datasheets/hiseq-3000-4000-specification-sheet-770-2014-057.pdf.

39. Duesberg PH, Vogt PK. Differences between the ribonucleic acids of transforming and nontransforming avian tumor viruses. Proc Natl Acad Sci U S A. 1970 Dec;67(4):1673–80.

40. Brugge JS, Erikson RL. Identification of a transformation-specific antigen induced by an avian sarcoma virus. Nature. 1977 Sep 22;269(5626):346–8.

41. Collett MS, Erikson RL. Protein kinase activity associated with the avian sarcoma virus src gene product. Proc Natl Acad Sci U S A. 1978 Apr;75(4):2021–4.

42. Levinson AD, Oppermann H, Levintow L, Varmus HE, Bishop JM. Evidence that the transforming gene of avian sarcoma virus encodes a protein kinase associated with a phosphoprotein. Cell. 1978 Oct;15(2):561–72.

43. Hunter T, Sefton BM. Transforming gene product of Rous sarcoma virus phosphorylates tyrosine. Proc Natl Acad Sci U S A. 1980 Mar;77(3):1311–5.

44. Stehelin D, Varmus HE, Bishop JM, Vogt PK. DNA related to the transforming gene(s) of avian sarcoma viruses is present in normal avian DNA. Nature. 1976 Mar;260(5547):170–3.

45. Czernilofsky AP, Levinson AD, Varmus HE, Bishop JM, Tischer E, Goodman HM. Nucleotide sequence of an avian sarcoma virus oncogene (src) and proposed amino acid sequence for gene product. Nature. 1980 Sep;287(5779):198–203.

46. Takeya T, Hanafusa H. Structure and sequence of the cellular gene homologous to the RSV src gene and the mechanism for generating the transforming virus. Cell. 1983 Mar;32(3):881–90.

47. Hanahan D, Weinberg RA. The hallmarks of cancer. Cell. 2000 Jan 7;100(1):57–70.

48. Nowell PC. Discovery of the Philadelphia chromosome: a personal perspective. J Clin Invest. 2007 Aug;117(8):2033–5.

49. Druker BJ, Tamura S, Buchdunger E, Ohno S, Segal GM, Fanning S, et al. Effects of a selective inhibitor of the Abl tyrosine kinase on the growth of Bcr-Abl positive cells. Nat Med. 1996 May;2(5):561–6.

50. Druker BJ, Talpaz M, Resta DJ, Peng B, Buchdunger E, Ford JM, et al. Efficacy and Safety of a Specific Inhibitor of the BCR-ABL Tyrosine Kinase in Chronic Myeloid Leukemia. N Engl J Med. 2001 Apr 5;344(14):1031–7.

51. Ley TJ, Mardis ER, Ding L, Fulton B, McLellan MD, Chen K, et al. DNA sequencing of a cytogenetically normal acute myeloid leukaemia genome. Nature. 2008 Nov;456(7218):66–72.

52. ClinVar [Internet]. Available from: `https://wwwlncbi.nlm.nih.gov/clinvar/`

53. dbGap [Internet]. Available from: `https://www.ncbi.nlm.nih.gov/gap`

54. Forbes SA, Bindal N, Bamford S, Cole C, Kok CY, Beare D, et al. COSMIC: mining complete cancer genomes in the Catalogue of Somatic Mutations in Cancer. Nucleic Acids Res. 2011 Jan;39(Database issue):D945-950.

55. Iorio F, Knijnenburg TA, Vis DJ, Bignell GR, Menden MP, Schubert M, et al. A Landscape of Pharmacogenomic Interactions in Cancer. Cell. 2016 Jul;166(3):740–54.

56. Wang T, Wei JJ, Sabatini DM, Lander ES. Genetic Screens in Human Cells Using the CRISPR-Cas9 System. Science. 2014 Jan 3;343(6166):80–4.

57. Behan FM, Iorio F, Picco G, Gonçalves E, Beaver CM, Migliardi G, et al. Prioritization of cancer therapeutic targets using CRISPR–Cas9 screens. Nature. 2019 Apr;568(7753):511–6.

58. Gao Y, Yan L, Huang Y, Liu F, Zhao Y, Cao L, et al. Structure of the RNA-dependent RNA polymerase from COVID-19 virus. Science. 2020 15;368(6492):779–82.

59. Lan J, Ge J, Yu J, Shan S, Zhou H, Fan S, et al. Structure of the SARS-CoV-2 spike receptor-binding domain bound to the ACE2 receptor. Nature. 2020;581(7807):215–20.

60. Protein Data Bank [Internet]. Available from: `https://www.rcsb.org/3d-view/6VSB/1`

61. Folding@home [Internet]. Available from: https://foldingathome .org/

62. Senior AW, Evans R, Jumper J, Kirkpatrick J, Sifre L, Green T, et al. Improved protein structure prediction using potentials from deep learning. Nature. 2020 Jan;577(7792):706–10.

63. Sinsheimer RL. The Santa Cruz Workshop- May - 1985. Genomics. 1989;4(4):954–6.

64. Botstein D, White RL, Skolnick M, Davis RW. Construction of a genetic linkage map in man using restriction fragment length polymorphisms. AM J Hum Genet. 1980;32(3):314–31.

65. Olson MV. Random-clone strategy for genomic restriction mapping in yeast. Proc Natl Acad Sci. 1986;83(20):7826–30.

66. Coulson A, Sulston J, Brenner S, Karn J. Toward a physical map of the genome of the nematode Caenorhabditis elegans. Proc Natl Acad Sci. 1986 Oct 1;83(20):7821–5.

67. Smith LM, Sanders JZ, Kaiser RJ, Hughes P, Dodd C, Connell CR, et al. Fluorescence detection in automated DNA sequence analysis. Nature. 1986 Jun;321(6071):674–9.

68. Watson JD, Berry A, Davies K. DNA: The story of the genetic revolution. New York: Knopf Publishing Group;

69. Kanigel R. The Genome Project. New York Times Magazine. 1987 Dec 13;

70. International Human Genome Sequencing Consortium. Initial sequencing and analysis of the human genome. Nature. 2001 Feb;409(6822):860–921.

71. Venter JC, Adams MD, Myers EW, Li PW, Mural RJ, Sutton GG, et al. The sequence of the human genome. Science. 2001 Feb 16;291(5507):1304–51.

72. Anderson S. Shotgun DNA sequencing using cloned DNAse I-generated fragments. Nucleic Acids Res. 1981;9(13):3015–27.

73. Fleischmann RD, Adams MD, White O, Clayton RA, Kirkness EF, Kerlavage AR, et al. Whole-genome random sequencing and assembly of Haemophilus influenzae Rd. Science. 1995 Jul 28;269(5223):496–512.

74. cshl.edu [Internet]. Available from: https://library.cshl.edu/ oralhistory/topic/genome-research/surprises-hgp

75. GigAssembler: an algorithm for the initial assembly of the human working draft. University of California, Santa Cruz; 2001. Report No.: UCSC-CRL-00-17.

76. Kent WJ, Sugnet CW, Furey TS, Roskin KM, Pringle TH, Zahler AM, et al. The human genome browser at UCSC. Genome Res. 2002 Jun;12(6):996–1006.

77. van Steensel B, Belmont AS. Lamina-Associated Domains: Links with Chromosome Architecture, Heterochromatin, and Gene Repression. Cell. 2017 May 18;169(5):780–91.

78. Amaral PP, Dinger ME, Mercer TR, Mattick JS. The eukaryotic genome as an RNA machine. Science. 2008 Mar 28;319(5871):1787–9.

79. Integrative Genome Viewer [Internet]. Available from: `https:// igv.org/app/`

80. Robinson JT, Thorvaldsdóttir H, Winckler W, Guttman M, Lander ES, Getz G, et al. Integrative genomics viewer. Nat Biotechnol. 2011 Jan;29(1):24–6.

81. UCSC genome browser [Internet]. Available from: `https://genome .ucsc.edu/`

82. Ensembl browser [Internet]. Available from: `https://www.sanger .ac.uk/too/ensembl-genome-browser/`

83. Yates AD, Achuthan P, Akanni W, Allen J, Allen J, Alvarez-Jarreta J, et al. Ensembl 2020. Nucleic Acids Res. 2020 08;48(D1):D682–8.

84. Reinert K, Langmead B, Weese D, Evers DJ. Alignment of Next-Generation Sequencing Reads. Annu Rev Genomics Hum Genet. 2015 Aug 24;16(1):133–51.

85. Marth GT, Korf I, Yandell MD, Yeh RT, Gu Z, Zakeri H, et al. A general approach to single-nucleotide polymorphism discovery. Nat Genet. 1999 Dec;23(4):452–6.

86. DePristo MA, Banks E, Poplin R, Garimella KV, Maguire JR, Hartl C, et al. A framework for variation discovery and genotyping using next-generation DNA sequencing data. Nat Genet. 2011 May;43(5):491–8.

87. Pfeiffer F, Gröber C, Blank M, Händler K, Beyer M, Schultze JL, et al. Systematic evaluation of error rates and causes in short samples in next-generation sequencing. Sci Rep. 2018 Jul 19;8(1):10950.

88. Cleary JG, Braithwaite R, Gaastra K, Hilbush BS, Inglis S, Irvine SA, et al. Joint variant and de novo mutation identification on pedigrees from high-throughput sequencing data. J Comput Biol J Comput Mol Cell Biol. 2014 Jun;21(6):405–19.

89. Aebersold R, Mann M. Mass-spectrometric exploration of proteome structure and function. Nature. 2016 15;537(7620):347–55.

90. Genome Aggregation Database Consortium, Karczewski KJ, Francioli LC, Tiao G, Cummings BB, Alföldi J, et al. The mutational constraint spectrum quantified from variation in 141,456 humans. Nature. 2020 May;581(7809):434–43.

91. Sutcliffe JG, Foye PE, Erlander MG, Hilbush BS, Bodzin LJ, Durham JT, et al. TOGA: an automated parsing technology for analyzing expression of nearly all genes. Proc Natl Acad Sci U S A. 2000 Feb 29;97(5):1976–81.

92. Lockhart DJ, Dong H, Byrne MC, Follettie MT, Gallo MV, Chee MS, et al. Expression monitoring by hybridization to high-density oligonucleotide arrays. Nat Biotechnol. 1996 Dec;14(13):1675–80.

93. Schena M, Shalon D, Davis RW, Brown PO. Quantitative monitoring of gene expression patterns with a complementary DNA microarray. Science. 1995 Oct 20;270(5235):467–70.

94. Stark R, Grzelak M, Hadfield J. RNA sequencing: the teenage years. Nat Rev Genet. 2019;20(11):631–56.

95. Stegle O, Teichmann SA, Marioni JC. Computational and analytical challenges in single-cell transcriptomics. Nat Rev Genet. 2015 Mar;16(3):133–45.

96. Price ND, Magis AT, Earls JC, Glusman G, Levy R, Lausted C, et al. A wellness study of 108 individuals using personal, dense, dynamic data clouds. Nat Biotechnol. 2017 Aug;35(8):747–56.

97. Poplin R, Chang P-C, Alexander D, Schwartz S, Colthurst T, Ku A, et al. A universal SNP and small-indel variant caller using deep neural networks. Nat Biotechnol. 2018;36(10):983–7.

98. Cho H, Wu DJ, Berger B. Secure genome-wide association analysis using multiparty computation. Nat Biotechnol. 2018;36(6): 547–51.

A New Era of Artificial Intelligence

We had dreams, Turing and I used to talk about the possibility of simulating entirely the human brain, could we really get a computer which would be the equivalent of the human brain or even a lot better?

Claude Shannon in *A Mind at Play: How Claude Shannon Invented the Information Age*

Artificial intelligence was born out of the dreams of its intellectual pioneers, foremost among them being Alan Turing, the godfather of AI and computing, and Claude Shannon, the founder of information theory. Their shared dream and separate works were found in Turing's *Computing Machinery and Intelligence*[1] published in 1950 and Shannon's *A Mathematical Theory of Communication*, published in 1948.[2] Turing, whose leadership and cryptography skills at Bletchley Park were key to breaking the German ENIGMA code during World War II, put forward his idea of a hypothetical computing machine in a seminal paper in 1936, and then he returned to designs of a computer after the war.[3] Turing's paper is considered by many in the field of computer science as one of the most important ever written, as he described the idea of universality in a computing device. The machine, now known as a *universal Turing machine*, could carry out any instruction or process that could be imagined. The breakthrough was Turing's definition of two new mathematical objects: a program and a machine. Thus was born the entirely new discipline of computer science, dominated ironically not by mathematicians, but by engineers.

Nearly in parallel and working on the other side of the Atlantic, Shannon was developing encryption systems for the Allied war effort. It was

Shannon's master's thesis from MIT, published in 1938, that created another cornerstone of modern computing, bringing together Boolean logic with circuit design.[4] The third foundational figure was John von Neumann, the founding father of game theory. He laid out some of the essential ideas for modern computer architecture and was an early proponent on modeling aspects of AI, based on information processing in the brain. The mathematical theories and conceptual brilliance of these innovators provided a rich milieu for AI researchers to work at the level of the building blocks or systems to produce intelligence in machines.

The development of the field of AI has been shaped largely by two divergent schools of thought as to how best to build problem-solving, thinking machines. Simply put, these are referred to as *top-down* and *bottom-up approaches*. The practitioners of top-down methods are *symbolists*; they apply reasoning and logic to model intelligent systems and use knowledge engineering to instruct machines. This is also known as *classical AI* or *symbolic AI*. This style dominated in the early phases of AI where machines could be programmed up front to solve mathematical theorems or robots could be taught to perform simple tasks. The first incarnations were blandly known as *knowledge-based systems*. Over the past several decades, a number of techniques have been developed for high-level reasoning tasks, the most formidable of which are a class of search algorithms that can evaluate all possible outcomes at a given decision point (lookahead search) and finite state machines that simulate sequential logic.

Adherents of the bottom-up camp start with simple components and utilize probability and statistics in a machine learning process conducted by a variety of different algorithms. At the dawn of AI, this was primarily done by using artificial neural networks inspired by brain architecture; these were the *connectionists*. Biological inspiration led to the first mathematical model of a single neuron, developed by McCulloch and Pitts in 1943.[5] In their model, a *binary threshold unit* (a simplified neuron) takes a number of inputs and computes a weighted sum followed by a nonlinear transform. By imposing a threshold, the unit is capable of implementing a linear discriminant. Responding to a pattern of continuous inputs with a single binary output, the threshold unit provides an intuitive bridge between the biological hardware of a spiking neuron and the task of categorization, a hallmark of cognition.

From early on, AI pioneers had the goal of teaching computers to do with their circuitry what humans could do with a brain, including the important capacities to reason, acquire, and integrate knowledge, and learn language, which developed only in the most recent epoch of primate evolution. Around 1950, Turing proposed a test to evaluate a machine's human-level abilities. The *Turing test* envisioned that human-level intelligence could be assessed by determining whether a machine could fool a human being in a conversation. In essence, if a person could not determine whether the dialogue they were conducting was with a machine or a human, then that was proof of human-level performance. Although this human-like AI definition has lost some of its luster over the ensuing decades, and the bar has been raised, the utility of benchmarking AI technologies against human capabilities remains entrenched in AI research labs and certainly in society at large.

Following the technological achievements coming out of World War II, and boosted by developments in neuroscience, psychology, and computing, there existed a naïveté that human-level machine intelligence could be achieved in the latter half of the twentieth century. Even at the turn of the twenty-first century, predictions continued to be made that *artificial general intelligence (AGI)* was just around the corner, perhaps within the decade. Ray Kurzweil, in his famously futuristic book, *The Singularity Is Near*, first published in 2005, puts the date for AGI (and a machine passing a Turing test) at around 2029.[6] Utilization of AGI is not proposed here in this book for biology and drug discovery; rather, task-oriented machine performance accomplished by "narrow AI" will be the source of AI's benefit to medicine and scientific progress for the foreseeable future.

AI Steps Out of the Bronx

Marvin Minsky and Frank Rosenblatt were the yin and yang of AI's roots, whose intertwined paths began at the Bronx High School of Science in New York City in the 1940s. The legendary high school was a cauldron of future innovation, with its former pupils leading science, technology, and engineering advances ranging from AI and the ARPAnet (Leonard Kleinrock, who was in Minsky's lab at MIT) to theoretical physics.[*] Minsky, who spent a lifetime in AI research, became the fiercest top-down advocate after spending time during his doctoral work contemplating neural networks. He took to modeling the function of the human brain with mathematics. He was also an inventor, having built a computing device called the SNARC, and he received a patent for the first design of a confocal microscope.[7] Rosenblatt came from a psychology background and developed a bottom-up connectionist approach for AI while at Cornell University. He published his one-layer neural network model known as the *perceptron* in 1958.[8] Rosenblatt was a natural marketer, and the machine was introduced to the public with much fanfare in the press. The device he constructed to implement the perceptron was called the *Mark I Perceptron*, which now resides at the Smithsonian Institution and is pictured in Figure 2.1.

Figure 2.1: The Mark I Perceptron
Used with permission from the Smithsonian Institution.

Part of Rosenblatt's research efforts in cognitive psychology was to see whether he could mimic the brain's ability to understand language and perform vision-based tasks like object classification. With the Mark I Perceptron, built not as a computer with Turing's universality in mind but as a device with a camera and a series of potentiometers (he had earlier written a perceptron-based program on an IBM 704 computer),

---

* The Bronx High School of Science has graduated an astonishing number of Nobel laureates (eight in total), including seven in physics: Leon Cooper (1972 Nobel prize) for BCS theory of superconductivity, Steven Weinberg and Sheldon Glashow (1979) for contributions to the electroweak theory, Melvin Schwartz (1988) for developing the neutrino beam method and elementary particle structure, Russell A. Hulse (1993) for co-discovery of pulsars, David Politzer (2004) for principles of quantum chromodynamics, and Roy Glauber (2005) for quantum theory of optical coherence.

he hoped to create a machine classifier. The military manual for the device detailed a protocol for classification experiments. He also anticipated the development of machines with human-like capabilities in the not-too-distant future. Rosenblatt's vision was remarkably bold but also of the era. A wildly optimistic view of technology's power had taken root in post-war America. Science fiction and the sci-fi movie genre that flourished in the early 1950s paralleled the birth of AI research and the atomic age. The epitome of this Pollyannaish view and of the technological naïveté found in the new AI research domain resided however with the symbolists. John McCarthy, who coined the term *artificial intelligence*, had boldly suggested along with other AI pioneers, including Minsky, that powerful capabilities from algorithms and computing machines were imminent.

"We propose a 2 month, 10 man study of artificial intelligence being carried out during the summer of 1956 at Dartmouth College in Hanover, New Hampshire. The study is to proceed on the basis of the conjecture that every aspect of learning or any other feature of intelligence can in principle be so precisely described that a machine can be made to simulate it. An attempt will be made to find how to make machines learn language, form abstractions and concepts, solve kinds of problems now reserved for humans, and improve themselves. We think that a significant advance can be made in one or more of these problems if a carefully selected group of scientists work on it for a summer."

John McCarthy, M.L Minsky, N. Rochester and C.E. Shannon,
A Proposal for the Dartmouth Summer Research Project on
Artificial Intelligence, August 31, 1955

Despite such hubris, McCarthy and Minsky were highly influential and together cofounded the MIT AI Project (which later became the MIT Artificial Intelligence Laboratory). Minsky's manifesto, *Steps Toward Artificial Intelligence,* published in 1961, made the case for symbol manipulation as the bedrock of AI and set out the path that AI researchers followed for more than half a century.[9] The following year, Minsky designed a Turing machine and proved it universal (a feat later turned by Stephen Wolfram for his machine in 2007). His laboratory seemed to be unleashing AI's promise, building graphical displays, simple robots, and further developing concepts of AI and theories of intelligence. He also crept into the public imagination as a technical advisor for Stanley Kubrick's masterpiece *2001: A Space Odyssey*, released in 1968. The following year, Minsky published *Perceptrons* with Seymour Papert, an influential and controversial book that attacked the work of Rosenblatt and delivered a death blow to single-layer neural networks.[10] In *Perceptrons*, the authors worked out the mathematics that demonstrated the limitations of connectionist learning machines. In fact, the book's front cover featured a maze-like square spiral, a geometric form that illustrated an example that cannot be captured by a simple neural network model. The perceptron's flaw turned out to be a missing piece of logic—the XOR gate. They also argued that feedforward networks of any size (that is, multilayered neural networks) could provide no guarantee of solving a problem optimally. Many in the field (then and now) had taken this to mean that neural networks were not useful; this was not the conclusion from their mathematics, as Minsky tried to point out over the years, but the damage was done.

The use of artificial neural networks as a problem-solving approach in AI and computer science—whether it be for pattern recognition, prediction, or learning new representations of input data—laid dormant during the period known as the first AI winter (1970 to the mid-1980s). The connectionists retrenched and continued to witness amazing developments across neuroscience, especially in the function and organization of the mammalian visual system, neuronal signal propagation and connectivity, and chemical basis of synaptic transmission. The explosive growth in computing power was beginning to shape what was possible in the laboratory and in industry. Perhaps, unexpectedly and most importantly, computational power and data storage drove an exponential growth in data. As if in universal synchrony, these trends would set the stage for a renaissance in neural network re-design and the emergence of machine learning as a major force shaping AI.

From Neurons and Cats Brains to Neural Networks

The structure of the mammalian brain and its cognitive capabilities have long served as inspiration and more recently as technical benchmarks for AI research. John von Neumann and Alan Turing, whose ideas shaped modern computing, both explored network models motivated by neural connectivity. The general principles of information processing, for vision in particular, have been extraordinarily useful for current AI technologies. However, the staggering complexity of a single neuron, much less networks with thousands of neurons and billions of synaptic connections, has kept a tight lid on solving the brain's ultimate mysteries. The profound lack of understanding of how humans can, for example, recognize the presence of a flower within tens of milliseconds after looking at a scene, or how and where the thought of a flower arises in consciousness, has served as fodder for skeptics to dismiss neurobiologically-based AI. The conclusion taken from this incomplete mechanistic understanding of the brain is that anything resembling neurocognition cannot be said to be truly modeled *in silico* or in mathematical models. Regardless, as nature has evolved an intelligent machine that may be incongruous in design (and not amenable to mass production in a factory), the route to many important insights in AI has come via modeling what sets of neurons do for processing sensory information utilizing probability theory and statistics—an approach that is fundamentally different from top-down logic and symbol manipulation.

Rosenblatt and the others who were building neural networks at the dawn of AI started with an abstract version of a simple neuron and gravitated toward modeling some narrow aspects of the visual system. Why start with processing visual information? The brain takes input from a sensory system that covers an array of sensing modalities—for light, taste, smell, sound, pain, thermal sensing, touch, balance, and proprioception. Receptors for the main senses are highly localized as for photoreceptors (in the eye), chemoreceptors (nose and mouth), mechanoreceptors (ear and skin), and the others are distributed widely throughout the body for proprioception, temperature sensing, and pain. Visual information carries a spatial component that can be captured only by a topographically organized visual field within the body's light sensor, the retina. Similar 3D maps for auditory signals and 2D maps for somatosensory perception are also critical. There is also an essential temporal component to almost all of these sensory data channels. Vision would seem most intuitive to study; perception and "seeing" are so entwined.

It is probably immediately obvious looking at the previous list that sensory data for vision would be easy to generate and interpret and has rich features and dimensionality that include contrasts, edges, depth, directionality, contours, and textures. Experimental data emerging from visual system neurophysiological studies could be used to update modeling strategies and to compare with machine output. The other advantage was that devices built for "computer vision" tasks would have research, military, and commercial applications. Working with light was relatively easy with optics and projection systems. A machine could be built with cameras serving as retinas, such as that which was attempted with Rosenblatt's Mark I Perceptron device.

The biological basis of vision is extraordinarily complex, but the main ideas driving neural network design and operation can be understood by learning the concepts of receptive fields, convergence, and feature detectors, along with tracing the hierarchical processing of sensory information from the retina to the primary visual cortex and finally forwarded onto higher cortical levels. Many or most of these pathways have feedback or reciprocal connections, which will be discussed later in the chapter. Although the early AI pioneers were aware of receptive field properties discovered in the visual system by Steven Kuffler, Horace Barlow, and Keffer Hartline, they did not have access to the hierarchical processing schemes and feature extraction capabilities of the brain; the first neural network models that arose in the late 1970s and 1980s did incorporate these essential concepts.

VISUAL INFORMATION PROCESSING IN THE RETINA

- In the mammalian visual system, photoreceptors lying at the back of the retina detect photons and transform the energy into an electrical signal as output.

- Parallel circuits in the retina further process the information, and the final output to the brain is delivered via a collection of 1.2 million retinal ganglion cells (RGCs) per eye (in humans).

- The RGCs deliver a train of spiking potentials that encodes the sensory information with spike frequency and other variables of the spike train.

- The receptive fields of the RGCs tile the entire visual field with a high degree of redundancy; there are more than 20 different types of RGCs receiving inputs from upstream parallel channels.

- RGCs can serve as pixel encoders, simply passing on raw pixel data to the brain, such as what occurs with cones in the fovea where there is a 1:1 correspondence of one cone to one RGC.

- The receptive fields of most RGCs are a result of convergence: a many-to-one mapping that produces a larger receptive field of the retinal image than individual photoreceptors. This convergence and expansion of receptive field continues into the cortex as representations of the data become more abstract and respond to large portions of the visual field.

■ RGCs are also feature detectors, with selectivity for edges and direction determined by center versus surround illumination properties built into their receptive fields.

■ Figure 2.2 shows examples of center-surround organization of RGC receptive fields.

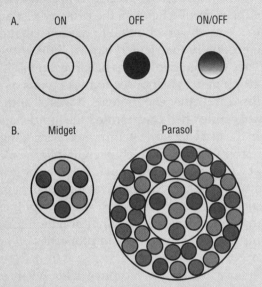

Figure 2.2: Receptive fields in retinal ganglion cells

A. Receptive fields of the various types of RGCs have a common organization defined by center-surround antagonism. The central core is excitatory, and the concentric surround region is typically inhibitory. ON RGCs respond to a small spot of light confined to the center area and fire action potentials; a larger spot crossing both the center and the surround regions weakens the response. OFF RGCs respond to a dark center and light surround; the ON/OFF type responds to both.

B. The receptive fields of the midget and parasol RGCs are shown at the bottom, with inputs from cones shown for illustration in shades of gray. In central retina (fovea), the receptive field center of a midget cell is a single cone, which drives color (wavelength) selectivity. By contrast, the larger parasol cells get a mixed cone input both in the center and the surround of their receptive fields.

In humans and other primates, more than 30 percent of the cortex is devoted to visual processing and higher-level perception tasks, which underscores its vital role in human behavior and survival. Visual information is sent to the brain from the retina's RGC axons, traveling down the optic nerve and relayed to neurons in a subcortical region known as the *lateral geniculate nucleus* (LGN). Neurons in this structure project out to a visual processing region in the back of the brain and known as area V1. This is the first stage of the cortical hierarchy for vision. At the level of the V1, there is a massive expansion of neural representation of the visual field, two orders of magnitude greater compared to the LGN inputs. By examining the receptive fields in the primary visual cortex (V1) of cats, Hubel and Wiesel were the first to discover that cortical neurons respond preferentially to oriented bars of light with edges, rather than to spots. They defined two classes of cortical cells, "simple" and "complex," based on neural responses to these types of simple visual stimuli. From these and other experiments, it was clear that local circuits within the visual cortex had transformed the input signals. By combining inputs from the local network, neurons were encoding orientation, directional selectivity (movement left or right into the receptive field), spatial frequency (that is, spacing of parallel lines or gratings), and binocularity. Critical for the subsequent development of neural network models like the neocognitron,[11] Hubel and Wiesel hypothesized a hierarchical organization for processing visual information to account for the increasing complexity of receptive field properties on neurons in the cat visual cortex.[12,13]

The emerging concept of a hierarchy of receptive fields, where increasingly more abstract fields are constructed from more elementary ones, led to searches for neurons that encoded final transformations into forms and signaled image recognition. Two higher regions have stood out. Complex motion and depth detection are localized to neurons in the middle temporal area (MT), while neurons in the inferotemporal (IT) cortex perform image (form, color) recognition. Work in a number of neurophysiology laboratories, starting in the late 1960s and perhaps most notably in Charles Gross' lab at Princeton, began to turn up neurons that appeared to respond only to faces, and very unique ones at that, like a grandmother.[14–16] Figure 2.3 shows a set of images used to elicit cortical neuron firing in the IT. The grandmother cell idea has had a long half-life, attracting equal amounts of fascination and ridicule. What would be the limit to the number of objects or images that the brain could store? The examples of extraordinary selectivity in image recognition have materialized in other structures, including the retina. A specific type of RGC in mice, known as W3, seems to respond only to pictures of a predator (for instance, a flying hawk) at a remarkably constant distance/time from a potential aerial attack.[17] It is now realized that such highly idiosyncratic feature detectors do exist, but it is unlikely that the sole function of these neurons is a set image (like your grandmother), as these can change over time and most neurons in the higher visual system are characteristically multifunctional.

Figure 2.3: Examples of shapes used to evaluate responses of neurons from the inferotemporal cortex

Recordings of neuronal activity within the inferotemporal cortex in awake primates revealed a surprising selectivity to a monkey-like hand image over any other hand shape or object. Animals were presented with each image or shape, and the responses were graded "I" for no response (the three images on left) to a maximum response of 6 for an image of a four-fingered hand on right. This was one of the first experiments suggesting that complex receptive fields existed in the visual hierarchy.

Figure source: Gross et al., Visual properties of neurons in inferotemporal cortex of the Macaque. J. Neurophysiology, vol. 35, page 104, 1972.

The cerebral cortex provides the brain with a massively parallel, biological computing architecture. All sensory modalities use parallel processing, and it is clear that evolution has converged on this efficient strategy to increase processing power. There is substantial evidence for multistage, hierarchically organized processing streams in all of the other sensory cortices, in addition to the visual cortical areas. A fairly uniform laminar structure comprised of six layers is shared throughout the extent of the cortex. A number of strategies appear to be utilized for connectivity within local modules and between structures in the visual processing hierarchy to achieve computational goals. Both excitatory and inhibitory neurons occupy distinct positions within the cortical structure; Figure 2.4 shows examples of excitatory pyramidal neurons residing in the mouse cortex.

The first modern connectivity map of the primate visual cortex was produced in 1991, and it revealed an elaborate visual information highway system whereby 30 visual areas were linked by 300 interconnections, as shown in Figure 2.5.[18] The diagram condensed years of mapping experiments, and it supported a hierarchical theme that was much more distributed in nature than first conceived by Hubel and Wiesel. A parallel, feedforward processing strategy could be inferred from the general design. Less appreciated were the reciprocal connections and potential feedback loops, their strengths, and ability to modulate processing in the network.

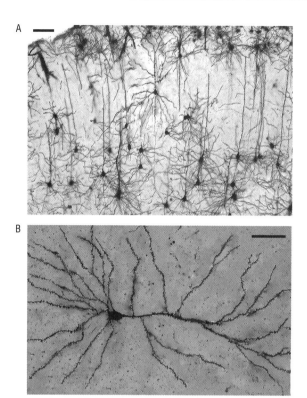

Figure 2.4: Pyramidal neurons in the cerebral cortex

A. Cross-section of a mouse brain with Golgi-stained pyramidal neurons in the cerebral cortex at 10X magnification (reference bar in upper-left corner is 100 microns). The staining reveals the laminar organization of the cortical structure, showing the positioning of pyramidal neurons across layers II through IV.

B. A Golgi-stained pyramidal neuron residing in layers II/III of the mouse cerebral cortex. The neuron's dendritic tree, cell body, and axonal projections can be easily seen at this magnification (black bar in upper-right is 50 microns). Dendritic spines are visible protruding along the entire length of dendritic processes, where interneuronal synaptic connections are established. Black dots distributed throughout the region are neuronal processes oriented perpendicular to the plane of this cortical neuron.

Images courtesy of Dr. Michael C. Wu, Founder and President of Neurodigitech, LLC in San Diego, California.

Replicating the feats accomplished in real time by the visual system with computer vision was a shared goal of the AI pioneers. Minsky's group at MIT in the 1960s had acquired a massive computer, and the robotics team figured that designing a rules-based program would endow a robot with human-like vision and motion control. The challenges of scenery interpretation and 3D motion were underappreciated, and ultimately the path to vision was not through symbolic manipulation. Likewise, Rosenblatt's perceptron machine and model could not accomplish the simple classification task for which it was designed, and his neural network idea and algorithms needed fundamental changes. Nonetheless, using information processing in the visual system as a guide led to essential breakthroughs. Neuroscience investigations suggested that a hierarchical processing scheme could enable a system to move from detection to feature extraction, providing the elements necessary for pattern recognition and ultimately categorization and image perception. The use of receptive field convergence and massively parallel computing strategies were a way forward for the next generation of neural network architectures that learned to "see."

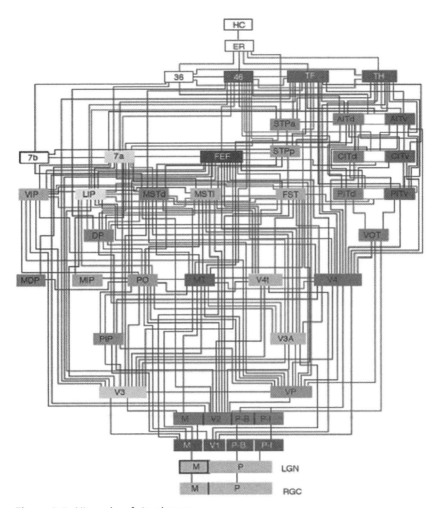

Figure 2.5: Hierarchy of visual areas

The classic wiring diagram of Felleman and Van Essen synthesized years of research on visual information processing and connectivity in the primate brain. Starting in the retina and moving into the LGC at the bottom of the diagram, connections are drawn to the primary visual areas such as the LGN, V1, V2, MT, and subregions of IT as discussed in the text.

From J. Felleman and D.C. Van Essen, "Distributed Hierarchical Processing in the Primate Visual Cortex," Cerebral Cortex 1991 1:1-47., by permission of Oxford University Press.

Machine Learning and the Deep Learning Breakthrough

"The only existence proof that any of the hard problems in artificial intelligence can be solved is the fact that, through evolution, nature has already solved them."

Terrence Sejnowski, *The Deep Learning Revolution*

The focus of AI in this book is on current and future applications in biology and medicine. AI as identified today is mostly *machine learning*. Data scientists and engineers in the field of machine learning design algorithms to process data, make predictions, and learn, borrowing and blending techniques from computer science, statistics, cognitive science, and other related disciplines. What makes machine learning so powerful for drug development projects is that modeling techniques can be imported from a wide variety of scientific fields and industrial applications without the need to re-invent anything.

Artificial neural networks are no different. The use of algorithms and neural networks in commercial applications first sprung up secretly in the financial industry in the late 1980s. Ever since then, algorithmic trading has had profound consequences on financial markets and the bottom lines of hedge funds, banks, and desk traders alike. By 2017, quant hedge funds accounted for 27 percent of all US stock trading. Machine learning tools were adopted quickly by large corporations in the 1990s to churn through large databases for credit analysis, fraud detection, and marketing. Prior to the set of advances that led to *deep learning*, a proliferation of machine learning methods was employed in industrial applications in business ranging from supply chain prediction to analysis of consumer behavior, presaging the recommendation systems later developed for consumers. From 1990–2010, the information revolution brought along further refinement of machine learning models, training paradigms, and increasingly powerful computational hardware in the form of graphics processing units (GPUs).

The deployment of ever more sophisticated machine learning approaches engineered to scale for massive datasets led to the takeover of retail by e-commerce giants such as Amazon and the mining of social network data for advertising by Google and Facebook. Deep learning artificial neural networks have already shown broad application across many other industries, and it is important to convey how these can address so many problems and why they work at a basic (nonmathematical) level. Several authoritative texts and scientific reviews that present AI and deep learning concepts, methods, and architectures in detail are listed in the "Recommended Reading" section at the end of this chapter.

Neural networks are computers, and they work magic as analysis tools. Modeling nonlinear phenomena (for example, weather, stock prices, drug-target interactions, and the multitude of other phase transitions found in chemistry and biology) is a core strength of neural networks. There is a remarkable range in design; some of the most successful neural networks have been loosely modeled on the architecture of networks operating in the brain. Processing information in artificial neural networks starts with simulating a highly interconnected structure with simple nodes, or neurons. Figure 2.6 shows the basic structure of a neural network (in this case a perceptron that lacks any hidden layers). In a fully connected, *feedforward* neural network with greater than one hidden layer, information flows in one direction, and the network is able to perform calculations that enable powerful classifiers to be built. The numerical value of a neuron at the input layer is simply the starting data (experimental data points, housing prices, and so forth).

In the diagram in Figure 2.6, inputs converge on a neuron at the output layer, and the input data is multiplied by the edge weights and summed to obtain the value of this neuron. Another important feature of the neural network is the use of an activation function, where the neuron's summed input is passed through a nonlinear function, such as tanh(x) or an S-shaped function like exp(x)/exp(x)+1, to obtain the value of the neuron. Thus, a neuron in an artificial neural network can operate like a simple-weighted classifier, with the output modified by a squashing function. More importantly, discriminating categories that are not linearly separable in the input require an intervening layer of nonlinear transformations between the input and output units of the neural network. This is analogous to a biological neuron's activity being a function of the sum of its neighboring presynaptic inputs. Although this network formulation vastly simplifies a neuron, the model is extremely useful, and modifying the synaptic inputs—the edge weights—is key to how learning is achieved in both.

As Minsky and Papert's *Perceptron* publication laid waste to the value of single-layer neural networks and induced the first "AI winter" across much of the field in the 1970s, symbolic AI continued forward as the dominant paradigm and led to the development of so-called *expert systems* in the 1970s–1980s. These systems were built on a framework where input data was evaluated with carefully designed rules, or heuristics and knowledge, that were coded in computer programs, which in turn executed tasks and generated answers. These expert systems handled specific tasks well but were brittle in the sense that they did not generalize—an expert system designed for one task could not cross terrain into a new problem area. Examples of these for biology and medicine were

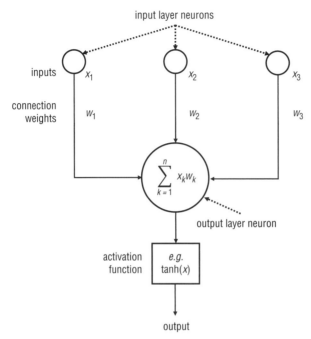

Figure 2.6: Information processing in neural networks

The basic structure of a neural network. The value of a neuron is specified by external data (inputs) or as a function of other upstream neurons in a feedforward network. The value of the downstream neuron (output layer neuron) is determined by multiplying the input values (x_k) with connection weights (w_k), which are summed, and an activation function (e.g. tanh(x)) applied to the temporary output. The final outputs can serve as inputs to the next layer of the network.

the INTERNIST system used for clinical diagnostic decision-making, Dendral for determining structures of organic compounds from mass spectrometry data, and the MOLGEN program for molecular biology, all developed at Stanford University. The definition of expert systems by one of the developers is enlightening:

> An "expert system" is an intelligent computer program that uses knowledge and inference procedures to solve problems that are difficult enough to require significant human expertise for their solution. The knowledge necessary to perform at such a level, plus the inference procedures used, can be thought of as a model of the expertise of the best practitioners in that field.
>
> *E.A. Fiegenbaum, Expert Systems in the 1980s*

During this time, an alternative approach was re-emerging whereby machines learned rules by experience to solve problems. The basic idea was that a system could be engineered to take input data, along with the answers (or labels), and the rules could be discovered by a learning algorithm. In this machine learning paradigm, a system is trained rather than programmed explicitly to perform this input to output mapping. The process of learning did not involve the symbolist's logic or hierarchical decision tree structures but rather employed statistical thinking. The methods of these new learning algorithms differed from statistics *per se* in that they dealt with large, complex datasets, with millions—if not billions—of data points. The various technical (philosophical) approaches taken to machine learning have led to what Pedro Domingos has called the five tribes of machine learning, presented in his remarkable book *The Master Algorithm* (in "Recommended Reading"). Table 2.1 highlights

Table 2.1: Five Approaches to Machine Learning

APPROACH	REPRESENTATION	ALGORITHMS	STRENGTH	REFERENCE
Analogizers	Support vectors	Kernel machines; support vector machines (SVMs)	Mapping to novelty	Hofstadter and Singer, *Surfaces and Essences* (2013)
Bayesians	Graphical models	Bayes' theorem; probabilistic programming; hidden Markov models (HMMs)	Uncertainty and probability	Pearl, *Probabilistic Reasoning in Intelligent Systems* (1988)
Connectionists	Neural networks	Backpropagation; gradient descent; LSTM	Parameter estimation	LeCun, Bengio, and Hinton, *Deep Learning Nature* (2015)
Evolutionists	Genetic programs	Genetic programming; evolutionary algorithms	Structure learning	Koza, *Genetic Programming* (1992)
Symbolists	Logic and symbols	Inductive logic programming; production rule systems; symbolic regression	Knowledge composition	Russell and Norvig, *Artificial Intelligence: A Modern Approach* (1995)

The table is based partly on a description of machine learning approaches that were characterized as the "five tribes" in *The Master Algorithm* by Pedro Domingos. His work (and Stuart Russell's) on a master algorithm hints at the possibility of combining and unifying some or all learning approaches, such as an integration of probability with logic.

the style, strengths, and algorithms typical of each approach. With the exception of the symbolic methods that do not necessarily need data to learn, all of the other modern approaches incorporate statistical machine learning on big data.

The network modelers and connectionists developed new ways of training feedforward multilayer networks with input-output pairs. The solution that was found for this problem is known as the *backpropagation algorithm* (or backprop), a stochastic gradient descent method that makes small, iterative adjustments to the edge weights between neurons to reduce the errors of the outputs.[19–22] It was "game on" for neural networks. The early developers in the PDP group in southern California showed that backprop could learn XOR; this important logic gate operation, so essential in circuits, can be implemented in a multilayered neural network. Undoubtedly, this influential algorithm led the entire field out of the AI winter in the mid-1980s and set the stage for deep learning, another generation later.

A second innovation came in the form of utilizing the hierarchical architecture found in the mammalian visual system for a new class of neural networks. These are known as *convolutional neural networks (CNNs)* and were first proposed in 1989 by Yann LeCun at the University of Toronto.[23] CNNs were among the first neural networks successfully trained for real-world applications (for example, LeCun's digit recognition program for ZIP codes). This was accomplished with the backprop algorithm, which was found to be more computationally efficient than other methods being tested at the time. Computer vision and image processing are problems readily tackled by CNNs, in addition to other continuous data with a grid-structured topology. This includes 1D audio, 2D images with three-channel (RGB) color, and certain types of 3D, volumetric imaging data, and video. Figure 2.7 presents a detailed dissection of the design of CNNs. LeCun's CNN architecture was designed to capture the feature detection and pooling strategy found in primate visual systems. It has achieved unparalleled success as a computer vision algorithm, and it has provided a basis for a computational theory of how the visual cortex works.

Another important domain of AI that saw resurgence in the 1980s was *reinforcement learning*. The concept of trial and error learning or learning by interaction with the environment is as old as computing, and the initial forays into AI were by psychologists and computer scientists. Claude Shannon developed a chess program to learn from trial and error. It was no less than Marvin Minsky who was interested in computational models of reinforcement learning and had laid out one of the core philosophical elements in his discussion of the credit assignment problem in his roadmap for AI.[9] Fast-forward a generation, and it was Richard Sutton's development of the temporal difference algorithm that launched reinforcement learning into the foreground of AI research in 1988.[24,25]

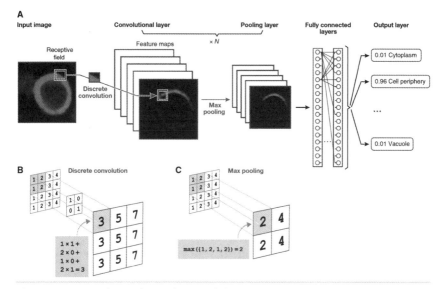

Figure 2.7: Convolutional neural network design concepts

A. Overview of training CNNs. Two concepts allow reducing the number of model parameters: local connectivity and parameter sharing. First, unlike in a fully connected network, each neuron within a feature map is only connected to a local patch of neurons in the previous layer, the so-called receptive field. Second, all neurons within a given feature map share the same parameters. Hence, all neurons within a feature map scan for the same feature in the previous layer but at different locations. The various feature maps might, for example, detect edges of different orientation in an image or sequence motifs in a genomic sequence.

B. The activity of a neuron is obtained by computing a discrete convolution of its receptive field, that is, computing the weighted sum of input neurons and applying an activation function. The convolutional kernel here is a 2X2 pixel matrix that slides along the image data, acting as filter to extract image features as in (A).

C. In most applications, the exact position and frequency of features are irrelevant for the final prediction, such as recognizing objects in an image. The pooling layer in effect summarizes adjacent neurons by computing, for example, the maximum or average over their activity, resulting in a smoother representation of feature activities. By applying the same pooling operation to small image patches that are shifted by more than one pixel, the input image is effectively downsampled, thereby further reducing the number of model parameters.

Figure reproduced and legend adapted from Angermueller, et al., (2016), "Deep learning for computational biology," Molecular Systems Biology 12:878, pp 1–16. Used under the terms of the Creative Commons Attribution 4.0 License.

Machine learning systems are built around the anticipated problem to solve, usually for some type of prediction, with careful consideration of the type of structured data available, the choice of algorithm, and the architecture employed to achieve the goals. Machine learning technologies can be categorized broadly based on how the learning algorithms are trained. Newer breeds of machine learning, including multitask, multimodal, transfer, and batch learning, will be taken up in the context of drug discovery and discussed in Chapters 6–8. A summary of the main styles of machine learning is as follows:

Supervised Learning   The goal of *supervised learning* is to learn by example. The strategy is to train the learning algorithm using test data with known targets or examples. The network builds a model by evaluating predictions with targets and adjusting weights to minimize an error or loss function. This instance is most widely used for classification and regression tasks, and the main applications in industry are found in this category, including image classification, language translation, and speech recognition.

Unsupervised Learning   The *unsupervised learning* style allows machine learning algorithms to discover patterns and features in the data without the aid of known targets. In contrast to supervised learning, the learning goal is not explicitly described. These methods are central to data analytics and are used in dimensionality reduction and clustering tasks.

Semisupervised Learning   As the name implies, *semisupervised learning* falls between the previous two categories, and it is used in applications where target labels are missing or unknown. The algorithms search for classes with limited information and learn to build boundaries between them. It is also employed with generative adversarial networks.

Reinforcement Learning   *Reinforcement learning* systems are built for sequential decision making, not for classification, as the learning influences a future action. The training is accomplished by a trial-and-error strategy following interaction with an environment. A reward function enables learning of good or bad choices in decision-making at each state (for example, in a game of chess, an algorithm can determine consequences of a move).

Deep Learning Arrives for AI

What deep learning has achieved in the short technological timeframe during which it has been around is truly astonishing. A cascade of events over the space of only a few years was needed to unleash the power of neural networks. First, a new algorithmic approach was needed to train deep neural networks. Geoff Hinton found the way with his Deep Belief Networks in 2006, which was the first important breakthrough.[26] On the computational side, the technology company NVIDIA released its CUDA development environment with an API in 2007, the earliest widely adopted programming model for GPU computing. The parallelism and multithreading capabilities of GPUs that provided the muscle for gaming were a perfect match for deep neural networks with millions of neurons and parameters to train.

Then, in 2011, the scientific community began writing CUDA implementations of neural networks. Andrew Ng at Stanford had teamed with NVIDIA and used 12 GPUs to match the computational output of 2,000 CPUs for deep learning. The final piece that fell into place and that continues to drive deep learning–based AI was the creation of massive datasets that could be used for training models.

In 2012, the AI field was electrified by the performance of AlexNet in the ImageNet challenge.[27] Using a clever approach to boost the training set of one million images, a new CNN architecture and a training program with GPUs, the Toronto team halved the error rate on the image recognition task compared to one year earlier. This jolt instantly made CNNs the field's primary computer vision algorithm and laid out a highly productive deep learning framework for research and industry. The framework consists of exploiting increased amounts of data and computation, combined with algorithmic improvements and new architectures built in such a way that they represent the structure in the data in a manner that is learnable with the selected algorithm.

The first significant commercial application of deep learning was the phonetic classification of speech to enable automated voice recognition. Natural language processing techniques of course existed well prior to deep learning. Many of these systems took over a decade to create and some (for instance, the CALO project, which led to Siri) were funded by DARPA and other agencies and large corporations. These R&D efforts eventually led to Apple's Siri, Google's Voice, Amazon's Alexa, Microsoft's

Cortana, and Baidu's Deep Speech 2. The technology giants were also building recommendation systems using natural language processing algorithms combined with other machine learning technologies. Training of computer and robotic vision systems with CNNs came next, with Google's Waymo division and Tesla leading the way in automotives. Outside of the technology bubble in Silicon Valley, any company with business decisions that could be tied to large-scale data was employing deep learning and using narrow AI's laser-like focus on a specific outcome to their competitive advantage.

In 2013, researchers at DeepMind published their work on DQN, an AI system using reinforcement learning and CNNs to play Atari video games. The innovative breakthrough was enough to convince skeptics that serious progress could be made on developing AI that generalizes. In addition, it was the first deep reinforcement learning application that worked reliably. The importance of the technology's promise and of the team's capabilities at DeepMind came to light at the beginning of 2014, when Google announced it was acquiring the UK company for an estimated $650 million. The same team went on to build AlphaGo and AlphaGo Zero that could best the world champion Lee Sodol at the complex strategy game Go. This 2017 event was watched by more than 280 million viewers in China, and it was later called China's Sputnik moment for AI technology by Kai-Fu Lee in his book *AI Superpowers*.[28] In biology and medicine, AI-based discoveries and platforms made significant advances beginning around 2015, achieving human-level (physician-trained) medical diagnostic accuracy from medical images (see the "Deep Learning's Beachhead on Medicine: Medical Imaging" section).

The initial foray of deep learning into drug development was far less dramatic. Artificial neural networks had their first heyday in molecular informatics and drug discovery approximately two decades ago. In 2013, public attention was drawn to a multiproblem QSAR machine-learning challenge in drug discovery posted by the pharmaceutical giant Merck. This competition on drug property and activity prediction was won by a deep learning network with a relative accuracy improvement of 14 percent over Merck's in-house systems and was reported in an article in *The New York Times*. While it is hard to pinpoint the first AI-based drug in development, the team at Insilico Medicine touted its initial success with GENTRL in discovering DDR1 kinase inhibitors as the frontrunner.[29]

The current obsession with deep learning by no means indicates that other AI progress has been stalled. Take, for example, the advances in strategy learning by Tuomas Sandholm from Carnegie Mellon University. Together with Noam Brown (CMU student at Facebook AI), the duo published another milestone in AI performance with a description of AI Pluribus in a paper in the journal *Science* in 2019. The publication was entitled "Superhuman AI for multi-player poker," which culminated years of work on the challenges of information-imperfect games where players are blind to cards the others are holding.[30] They developed an interacting set of algorithms to evaluate a poker-playing strategy, which was trained entirely by self-play.

The challenges in complex strategy games, especially with those involving multiple players, are directly related to randomness and imperfect information and the high computational costs involved in searching for all possible outcomes and optimal actions. AI Pluribus' technical innovations were laid out in the paper as falling into three categories:

- A depth-limited search algorithm that reduced computational resources by an estimated five orders of magnitude

- An equilibrium finding algorithm that improved on the standard Monte Carlo Counterfactual Regret (MCCFR) with a linear MCCFR approach

- Additional algorithmic shortcuts to improve memory usage

The blueprint computation algorithm learns strategy by self-play, and a real-time strategy is computed with the aid of the innovations developed.

Deep Neural Network Architectures

At the heart of creating an AI-based application, or configuring an AI-driven approach to analyze a dataset, is the underlying architecture. Figure 2.8 shows the set of the most commonly deployed neural network architectures with a short description.

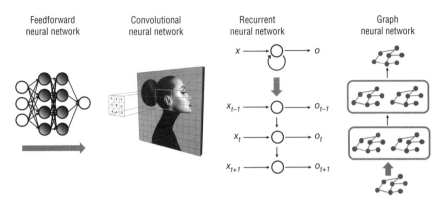

Figure 2.8: Neural network architectures

Graphical representation of the major neural network architectures. Feedforward networks pass inputs into the network in one direction with hidden layers fully connected (see Figure 2.6). Convolutional neural networks are designed to process information from grid-like topologies utilizing kernel convolutions and pooling (see Figure 2.7). An example of one recurrent neural network architecture (there are many) is shown. RNNs use self-loops where neurons get inputs from other neurons at previous time steps (*t-1*, *t*, and *t+1* in the example). Graph neural networks such as the one depicted utilize hidden layers and processes to handle unstructured data in graph models.

There are infinite possible configurations of networks and more complex systems that are made by varying the neuron layers by depth and width, plugging modular combinations together (for example, a CNN plus an RNN), and finding combinations of learning and search algorithms that are particularly productive for a given task. The AlexNet CNN had five convolutional layers in series, followed by three successive, fully connected layers to improve image classification. On the other end of the extreme, one group created a CNN for a medical classification task that had 172 convolutional layers. In the Atari game-playing agent built by DeepMind, AlexNet's five-layer CNN was employed as its perception component and used a search algorithm to determine the best moves, which then communicated with the joystick movement controller. The novel architecture for AlphaGo Zero was built to allow self-play of Go (4.9 million games) to train the model in an unsupervised fashion by reinforcement learning with a neural network that incorporated a Monte Carlo Tree Search lookahead search algorithm into a neural network consisting of "many residual blocks of convolutional layers."[31]

The Chinese company Baidu's speech recognition system, Deep Speech 2, is a perfect example of a modular architecture. The problem is challenging—take an audio recording as input to the network and output a written transcription. Recurrent neural networks are great for text sequence, but first the system must process the audio. The way that Baidu engineers addressed this was to employ a modular design. The input data was an audio spectrogram—essentially a black-and-white image conveying the intensity of frequencies detected in the speech pattern. The audio is converted into pixels, and convolutional layers take the input to extract features by CNN magic. The output feeds into a large number of RNN layers, which extracts the word sequence relationships. Next, the RNN connects to a series of fully connected layers. The next processing stage uses a new technique to search for alignment between the label and the fully connected layers. To aid in error correction, the system has access to the frequencies of words and their use in the language. The final process takes the predictions from the network and the language statistics to make the most probable guesses as to the phrase structure. For example, if the system was examining the phrase "the burglar took honey at the teller window," it could recognize the misuse or mislabeling of the word *honey* and even supply a missing word based on statistical frequencies.

The application of AI and neural network architectures to drug discovery and development is still in its infancy. To look ahead, a more detailed discussion of advances in RNN architectures and the reinforcement learning paradigm will be presented in Chapter 5, "Healthcare and the Entrance of the Technology Titans." These systems may turn out to be important in digital health applications and novel biological and nonbiological therapeutic intervention strategies for behavioral and neuropsychiatric diseases. The use of long short-term memory (LSTM) and generative adversarial networks (GANs) in medical imaging, genomics, and areas of drug discovery will be found in Chapter 7 and Chapter 8.

Evaluating complex, dynamic systems that are typical of biology and medicine might require co-opting architectures that have been constructed for robotic control and self-driving cars. The industry has converged on three-layer architectures for control of self-driving vehicles, conceived by teams in the DARPA Grand Challenges of the early 2000s. The models incorporate context with planning in real time, with a sensing layer for processing sensor data, a perception layer for understanding terrain, and a reasoning layer. The latter incorporates a route planner, a "state" evaluator, and a motion planner that does the thinking for the navigation tasks based on the current and foreseeable situation. In the not-too-distant future, drug and medical exploration activities might want to build a system that models drug *pharmacodynamics*, the physiological changes of an organ system over time, or model all of the potential avenues available for regulating expression of genes to control disease processes. What might be accomplished with AI and biological data is likely limited only by the pharmaceutical industry's imagination.

Deep Learning's Beachhead on Medicine: Medical Imaging

Healthcare undoubtedly will be one of the early winners of the AI era. One of the major premises of this book is that AI technology will transform the future of medicine by unlocking advances in biology that lead to novel therapeutic development and secondarily through enhancing technologies central to healthcare. AI is already showing great promise for clinical diagnostic procedures, led most recently by the explosion of applications using deep learning for medical image analysis. This is where the AI technology is storming the beach of medicine.

The importance of diagnostic imaging for healthcare is underscored by the fact the US healthcare system produced more than 800 million medical images in 2017. The breakthrough in CNN performance by AlexNet on image classification tasks in 2012 was a harbinger of changes to come in healthcare and other industries where image data has significant commercial or scientific relevance. The technology sector and e-commerce companies were first to capitalize on the advances in deep learning from CNNs, and it wasn't until presentations at workshops and conferences in 2013–2014, followed by the first journal papers in 2015, did CNNs begin to make their way from R&D into medical imaging with potential diagnostic applications. Venture funding of medical AI imaging startups began around 2014; the rush into AI-based algorithms and startups will be explored in depth in Chapter 6, "AI-Based Algorithms in Biology and Medicine." Deep learning techniques have now basically taken over the field of medical image analysis, with CNN architectures dominant.

The integration of *computer-assisted diagnosis (CAD)* into clinical practice emerged as a goal in medicine more than 60 years ago, coincident with the dawn of computers, AI, and statistical machine learning. What took so long? It was recognized early on that statistical methods could be developed to assist with diagnosis. Ideas were proposed to use symbolic logic systems to perform differential diagnosis (1959), Bayes' theorem to predict tumor type,[32] and "optical scanning and computer analysis" as an aid for mammography (1967). These methods never entered the real-world, clinical mainstream, and they remained largely confined to academic medicine. Once radiographic images, such as those from chest X-rays, could be encoded (digitized) and analyzed by computers starting in the 1970s, a more serious move toward CAD and digital image analysis was warranted.

Two additional inventions in the 1970s led to a revolution in computer-based diagnostic medicine: computerized axial tomography (CT or CAT scanners) and magnetic resonance imaging (MRI) technology. These new imaging modalities had the power to visualize soft tissue and 3D structure with remarkable resolution. The first images for medical MRIs were obtained with 0.5T magnets in 1977. The most recently developed 7.0T MRI machines will enable viewing cells at a submillimeter scale and potentially open a window on neuronal circuit activity in the human cortex.[33]

One of the ingredients for the progress of CAD was the creation of *Picture Archive and Communications Systems (PACS)*, a term coined in 1982, an effort that required arduous integration of clinical, radiology department, and hospital information management systems. However, even with carefully designed expert systems and the adoption of PACS in the twenty-first century, the role of CAD in the clinic remained relatively minor, as these systems were not good enough to replace trained radiologists and added to their workloads.

The challenge for CAD and human-level performance cuts across four areas of medicine with strong dependence on images in clinical workflows: radiology, pathology, dermatology, and ophthalmology. The long sought-after destination is an AI component that could improve efficiency for the diagnostic phase of healthcare by minimizing errors, reducing workloads, and allowing radiologists to work with pre-analyzed and identified features in complex scans. Across workflows in diagnostic medicine utilizing x-rays, MRI, CT, and PET scans, imaging analysis requires strong performance in feature extraction, pattern recognition, and classification. For example, image-based tasks in clinical oncology are generally separated into detection of tissue abnormalities (cancerous nodules, lumps, calcifications, and so on), characterization (boundary definitions by segmentation algorithms, diagnostic classification, and tumor staging), and, in many cases, post-diagnostic monitoring to track temporal changes in response to therapies. Computational pipelines developed by engineering or "handcrafting" features of relevance to these specific tasks provide a means of quantification. The predefined, disease-related features, such as a tumor's 3D shape or texture, are used as inputs to train machine learning models, such as support vector machines or random forests, which then classify tumors or patients into categories to provide clinical decision support. Nearly all of these image-based radiology tasks are contingent upon the quantification and assessment of radiographic characteristics from images.

The research community has now shown that almost the entire spectrum of imaging tasks is accomplished with greater wizardry by deep learning architectures than perhaps by any other machine learning approach. Well over 100 clinical studies have been performed with a variety of deep neural network architectures for image segmentation, organ detection, and disease characterization tasks. A great illustration of the methodologies and improvements brought by CNNs for cancer diagnosis from images comes from work by Kun-Hsing Yu and colleagues. In the first publication from 2016, Yu in Michael Snyder's group at Stanford tapped into a large archive of histopathology slides available from TCGA (see Chapter 1) for analysis of two subtypes of nonsmall cell lung cancer.[34] They designed a pipeline to capture images and classify the main subtypes, adenocarcinoma and squamous cell carcinoma, which together account for more than 80 percent of lung cancers worldwide. Using a software package known as CellProfiler, nearly 10,000 features were extracted. A subset of features was subsequently selected and used as input into a variety of machine learning algorithms. On the classification task, the data revealed that the AI-based system had comparable performance to trained oncologists on predicting type and severity of lung cancer from histopathology slides. Yu then went on to build CNNs that could be used to extract features without the CellProfiler steps; Table 2.2 compares the performance between all of the algorithms.[35]

How have CNNs on tasks within medical imaging achieved or surpassed human-level performance and become so favored? The primary reason that CNNs are winning against alternative machine learning approaches is due in large part to the battle being won between data—the heart and soul of training deep neural networks—and domain expertise. With big data, deep neural networks—particularly CNNs—can automatically learn feature representation and identify abstract features beyond what any shallow neural network or machine learning algorithm can extract. The availability of rapidly accumulating amounts of imaging data residing in PACS only adds to the advantage. Another technical edge is that CNNs can generalize better than hand-crafted features when asked to perform similar tasks on different datasets or other imaging modalities. The deep learning approach is also robust against undesired variability in the form of inter-user differences in feature selection. Finally, a formidable practical advantage for CNNs is their built-in, automated feature learning, which eliminates the manual

Table 2.2: Performance Comparison of Machine Learning Methods for Lung Cancer Diagnosis in the TCGA Test Set

MACHINE LEARNING METHOD	LUAD VS. DENSE BENIGN TISSUE	LUSC VS. DENSE BENIGN TISSUE	LUAD VS. LUSC
Convolutional neural network-based			
AlexNet	0.95	0.97	0.89
GoogLeNet	0.97	0.98	0.90
VGGNet	0.97	0.98	0.93
ResNet	0.95	0.94	0.88
Non-neural network-based			
SVM with Gaussian kernel	0.85	0.88	0.75
SVM with linear kernel	0.82	0.86	0.70
SVM with polynomial kernel	0.77	0.84	0.74
Naïve Bayes classifiers	0.73	0.77	0.63
Bagging	0.83	0.87	0.74
Random forest using conditional inference trees	0.85	0.87	0.73
Breiman's random forest	0.85	0.87	0.75

Areas under the receiver operating characteristics curves (AUCs) of the classifiers for distinguishing lung adenocarcinoma (LUAD) from adjacent dense benign tissues, lung squamous cell carcinoma (LUSC) from adjacent dense benign tissues, and LUAD from LUSC are shown. Adapted from Yu et al., "Classifying non-small cell lung cancer types and transcriptomic subtypes using convolutional neural networks," Journal of the American Medical Informatics Association, 27(5), 2020, 757–769, by permission of Oxford University Press.

preprocessing steps of other machine learning and CAD processes in the imaging workflow.

The first indications that CNNs can attain human-level performance was reported for MRI-based prostate cancer diagnosis in 2015.[36] One of the most extraordinary, and potentially most significant for worldwide health, was the demonstration of expert-level chest X-ray diagnosis by deep learning from another Stanford team led by AI guru Andrew Ng.[37] The researchers developed a CNN called CheXNeXt and trained a model on more than 100,000 chest X-rays found in the chestX-ray14 database. The trained algorithm achieved better performance on the pneumonia detection task compared to a panel of expert radiologists. The global burden of tuberculosis could be reduced by using AI to analyze thoracic scans for millions of at-risk people in an effort to discover and reduce spread of disease in less-developed countries. Numerous studies are now documenting impressive performance of deep learning algorithms for medicine, with breakthrough results in mammography, diabetic retinopathy, and malignant dermatologic lesions.

How will AI eventually be implemented for safe medical practice? There is good evidence that deep learning systems can surmount the hurdle of clinical utility, whereas earlier CAD/feature engineering methods have failed. One potential future for AI in medicine is first to augment a pathologist's subjectivity in diagnosis with tumor, cell, and molecular-level quantitative measurements. Even with limited integration into healthcare, a well-engineered and automated approach with a physician in the loop will lower costs and greatly improve patient outcomes. Overcoming challenges in data collection, model training, and clinical deployment will lead to incorporation of best-in-class algorithms using deep learning architectures into safe and proven domains, as shown in Figure 2.9.

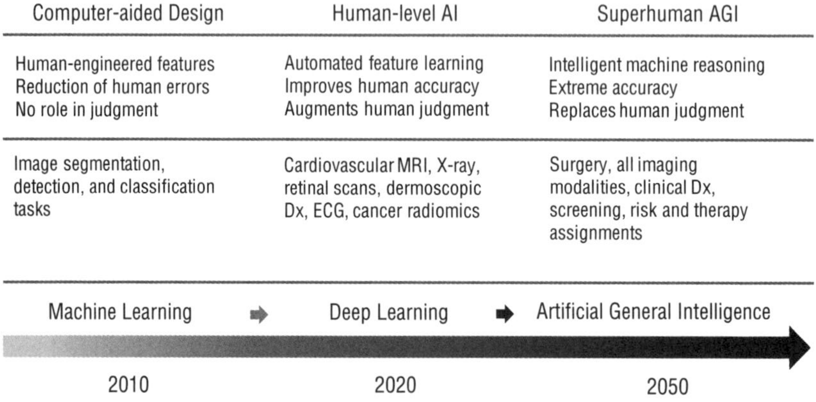

Computer-aided Design	Human-level AI	Superhuman AGI
Human-engineered features Reduction of human errors No role in judgment	Automated feature learning Improves human accuracy Augments human judgment	Intelligent machine reasoning Extreme accuracy Replaces human judgment
Image segmentation, detection, and classification tasks	Cardiovascular MRI, X-ray, retinal scans, dermoscopic Dx, ECG, cancer radiomics	Surgery, all imaging modalities, clinical Dx, screening, risk and therapy assignments
Machine Learning ➡	Deep Learning ➡	Artificial General Intelligence
2010	2020	2050

Figure 2.9: A future path for integrating AI into medical practice

The diagram maps a hypothetical progression from CAD in 2010, human-level AI around 2025, and then superhuman AGI in 2050. Some of the medical and diagnostic procedures anticipated to be performed by machines are indicated.

Limitations on Artificial Intelligence

Many prominent observers, AI researchers, and scientists are quick to call out the limitations, denounce the hype, and express skepticism around the depth of progress that has been made toward building full-fledged AI. Even further, the AI pioneer's dream of intelligent machines that can reason and grasp abstract concepts like schadenfreude, human rights, or injustice continues to be pushed far out in the future. Part of this is a counter-reaction to the media, which is looking for proof of superhuman intelligence today and the next big thing in the digital realm tomorrow. If the yardstick used to measure AI is the degree to which any system fully recapitulates human cognitive functions or capacities (for example, language, creativity, vision, or emotional intelligence), there will be disappointment. We are nowhere near an understanding of how the brain works. Neuroscientists who have made attempts to develop theories of mind out of connectionism, and those promoting efforts like the Human Brain Project, admit that there is a long way to go. The early, lofty goal of AI and brain researchers that we might come to explain consciousness or produce it in machines once we develop suitable models and computing power has been largely abandoned for now.

In AI, progress had flatlined for decades; only recent technological developments have led to mind-blowing commercial and scientific applications. The startling progress made by deep learning points to the fact that AI developments are tightly linked to technology and tools, and not limited so much by a need to surface insights continually through discovery, as required in many scientific disciplines. AI is an engineering field, although it does share some commonalities with biology and physics around experimental design and hypothesis testing. As outlined in Chapter 1, for biology, it is equally the case that computational power has enabled AI to reach some stunning plateaus with human-level and even superhuman performance.

What are the limits of AI and to what can they be attributed? Within the AI community, the criticisms fall into two areas: the first being leveled at attempts to build AGI and the other at limitations specifically of the deep learning paradigm (and the use of deep neural networks in building components of AGI). Some of these are listed here:

- AGI will not be achieved with approaches that do not have the design features incorporated for reasoning and planning, a specific criticism of artificial neural networks and narrow AI.

- Current approaches toward AGI (such as at OpenAI's GPT-3 language model) lack an ability to understand context. This limits understanding of meaning in speech, time-series data, and other interpretation of scenes.

- Theoretical progress into understanding deep learning is poor; making intelligent machines without a theoretical grounding will lead to development roadblocks.

- Deep learning does a poor job at modeling the brain and is not the way forward, as humans learn with far less data.

- Deep learning fails to generalize. Trained models can't learn beyond boundaries easily.

- Deep learning struggles with the overfitting problem; deep neural networks are trained to memorize examples.

- Current deep learning methods are recycled human intelligence—really nothing more than a better pattern recognition algorithm.

- There is evidence that deep learning performance has leveled off, signaling a potential dead-end to deep learning approaches for AGI.

For the pharmaceutical industry, the concerns are more practical and industry-specific:

- Data requirements limit application of deep learning: Pharma companies indicate that structured data is limited in availability, acquiring such data is expensive, and the methods by which the data were obtained often render it unreliable.

- Data sharing barriers due to competition make it difficult to amass important datasets. Medical privacy limits data access to critical medical records information for clinical trial analysis.

- Application domains are limited for deep learning in drug development (mainly virtual screening, QSAR).

- Efficient machine learning algorithms already exist and do not require deep learning approaches or large datasets.

- AI offers only marginal improvement. Small increases in yield or efficiency with deep learning aren't worth the investment in data science, computer science, and high-performance computing.

- Interpretability. Deep learning algorithms are black boxes with limited understanding of how results were obtained and why predictions were made.

At the societal level, there are many concerns that are well-known and being actively debated, most having to do with AI bias (gender, race, class, and health status discrimination), job loss and the social disruption it creates, and human loss of decision-making autonomy. Across society, a major fear on the far-reaching capabilities of AI has been that many jobs will be lost to algorithms and computers and machines that can replace human skills and hundreds, if not thousands, of humans doing the same tasks. In healthcare, AI technology would be extraordinarily well-suited to diagnostics, which at its core takes a collection of data from a patient (symptoms, observations, scans, medical history, molecular profiling data, genetics, socioeconomic data, and so forth) and predicts some outcome, recommends an action, or infers a cause. Outcomes would include a pathophysiological state (an illness), a probable cause of illness, and recommendations for therapies, secondary diagnostic procedures, or any other medically valuable action for an individual. Searching for correlations in high-dimensional data and making predictions are precisely what deep learning is built to excel at.

In the medical professional community, there has been a fear that AI-based diagnostics might replace radiologists entirely. The futurist Yuval Harari suggested that AI is so well-suited to the medical diagnostic problem that the eventual development of a medical robot armed with sufficient training, vast knowledge, and accurate predictive capabilities with superhuman performance would make physicians and physician training obsolete. The current thinking of most experts at the interface of AI and medicine believe that AI will not replace physicians anytime soon. As Eric Topol elegantly laid out in his book *Deep Medicine*, the role of AI will be to augment physicians' capabilities, providing them with expert second opinions and also saving vast amounts of time, allowing doctors to spend more time with patients.[38]

Another enormous worry has taken root as technology companies such as Facebook, Google, Microsoft, Amazon, and Apple in the United States and Baidu and Tencent in China have enormous technical resources, market power, and AI capabilities that are global in reach and scale. How will they operate AI for society's benefit? Michael I. Jordan, a prominent AI scholar and professor at the University of California, Berkeley, has warned that society's input and a new engineering discipline is needed for the design of "societal-scale decision-making systems." AI researchers in academia and industry have assembled at summits to address the fears and moral issues around AI: one of the first was organized by Max Tegmark of MIT that took place in Puerto Rico in 2015. Another meeting was then held on the Asilomar conference grounds in Pacific Grove,

California in 2017, echoing the importance of considering technology's impact—as recombinant DNA researchers did at the same site in the 1970s.

Artificial intelligence has entered an era in which computational techniques directly inspired by the brain will continue to power data-driven applications across many industries. With commercial incentives, open frameworks, computational power, and algorithm tweaks will come major progress. The tech industry up to this point has driven these advances and the underlying engineering, a point yet to be learned by the pharmaceutical industry.

Notes

1. Turing AM. Computing Machinery and Intelligence. Mind. 1950;59(236):433–60.

2. Shannon CE. A Mathematical Theory of Communication. Bell Syst Tech J July Oct 1948.

3. Turing AM. On computable numbers, with an application to the Entscheidungs problem. Proc Lond Math Soc. 2(1936):230–65.

4. Shannon CE. A Symbolic Analysis of Relay and Switching Circuits. Trans Am Inst Electr Eng. 1938;57:471–95.

5. McCulloch WS, Pitts W. A logical calculus of the ideas immanent in nervous activity. Bull Math Biophys. 1943;5:115–33.

6. Kurzweil R. The Singularity is Near: when humans transcend biology. New York: Viking Press; 2005.

7. Minsky M. Microscopy Apparatus. 3,013,467, 1961.

8. Rosenblatt F. The perceptron: a probabilistic model for information storage and organization in the brain. Psychol Rev. 1958;65 (6):386–408.

9. Minsky M. Steps Toward Artificial Intelligence. Proc IRE. 1961 Jan;49(1):8–30.

10. Minsky M, Papert S. Perceptrons. Cambridge, MA: MIT Press; 1969.

11. Fukushima K. Neocognitron: A self-organizing neural network model for a mechanism of pattern recognition unaffected by shift in position. Biol Cybern. 1980 Apr;36(4):193–202.

12. Hubel DH, Wiesel TN. Receptive fields, binocular interaction and functional architecture in the cat's visual cortex. J Physiol. 1962 Jan;160:106–54.

13. Hubel DH, Wiesel TN. Receptive fields and functional architecture in two non-striate visual areas (18 and 19) of the cat. J Neurophysiol. 1965;18:229–89.

14. Gross CG, Bender DB, Rocha-Miranda CE. Visual Receptive Fields of Neurons in Inferotemporal Cortex of the Monkey. Science. 1969 Dec 5;166(3910):1303–6.

15. Gross CG, Rocha-Miranda CE, Bender DB. Visual properties of neurons in inferotemporal cortex of the Macaque. J Neurophysiol. 1972 Jan;35(1):96–111.

16. Desimone R, Albright TD, Gross CG, Bruce C. Stimulus-selective properties of inferior temporal neurons in the macaque. J Neurosci. 1984;4(8):2051–62.

17. Zhang Y, Kim I-J, Sanes JR, Meister M. The most numerous ganglion cell type of the mouse retina is a selective feature detector. Proc Natl Acad Sci. 2012 Sep 4;109(36):E2391–8.

18. Felleman DJ, Van Essen DC. Distributed Hierarchical Processing in the Primate Cerebral Cortex. Cereb Cortex. 1991 Jan 1;1(1):1–47.

19. Werbos P. Beyond regression: New tools for prediction and analysis in the behavioral sciences. Ph.D. dissertation, Committee on Appl. Math., Harvard University; 1974.

20. Werbos PJ. Backpropagation through time: what it does and how to do it. Proc IEEE. 1990 Oct;78(10):1550–60.

21. Rumelhart DE, Hinton G, Williams RJ. Learning internal representations by error propagation. In: Parallel Distributed Processing: Explorations in the microstructure of cognition. MIT Press; 1985. p. 318–62.

22. Hinton GE, McClelland JL, Rumelhart DE. Distributed representations. In: Parallel Distributed Processing: Explorations in the microstructure of cognitions. MIT Press; 1986. p. 77–109.

23. LeCun Y. Generalization and network design strategies. Technical Report CRG-TR-89-4. University of Toronto Connectionist Research Group; 1989.

24. Sutton, RS. Learning to predict by the methods of temporal differences. Mach Learn. 1988;3:9–44.

25. Sutton, RS, Barto, AG. Reinforcement Learning: An Introduction. Second. The MIT Press; 2018.

26. Hinton GE, Osindero S, Teh Y-W. A Fast Learning Algorithm for Deep Belief Nets. Neural Comput. 2006 Jul;18(7):1527–54.

27. Krizhevsky A, Sutskever I, Hinton GE. ImageNet Classification with Deep Convolutional Neural Networks. In: Pereira F, Burges CJC, Bottou L, Weinberger KQ, editors. Advances in Neural Information Processing Systems 25. 2012. p. 1097–1105.

28. Lee K-F. AI Superpowers: China, Silicon Valley and the new world order. Boston: Houghton Mifflin Harcourt; 2018.

29. Zhavoronkov A, Ivanenkov YA, Aliper A, Veselov MS, Aladinskiy VA, Aladinskaya AV, et al. Deep learning enables rapid identification of potent DDR1 kinase inhibitors. Nat Biotechnol. 2019;37(9): 1038–40.

30. Brown N, Sandholm T. Superhuman AI for multiplayer poker. Science. 2019 Aug 30;365(6456):885–90.

31. Silver D, Schrittwieser J, Simonyan K, Antonoglou I, Huang A, Guez A, et al. Mastering the game of Go without human knowledge. Nature. 2017 Oct;550(7676):354–9.

32. Lodwick GS, Haun CL, Smith WE, Keller RF, Robertson ED. Computer Diagnosis of Primary Bone Tumors. Radiology. 1963 Feb 1;80(2):273–5.

33. Nowogrodzki A. The world's strongest MRI machines are pushing human imaging to new limits. Nature. 2018 Oct 31;563(7729): 24–6.

34. Yu K-H, Zhang C, Berry GJ, Altman RB, Ré C, Rubin DL, et al. Predicting non-small cell lung cancer prognosis by fully automated microscopic pathology image features. Nat Commun. 2016 16;7:12474.

35. Yu K-H, Wang F, Berry GJ, Ré C, Altman RB, Snyder M, et al. Classifying non-small cell lung cancer types and transcriptomic subtypes using convolutional neural networks. J Am Med Inform Assoc. 2020 May 1;27(5):757–69.

36. Litjens GJS, Barentsz JO, Karssemeijer N, Huisman HJ. Clinical evaluation of a computer-aided diagnosis system for determining

cancer aggressiveness in prostate MRI. Eur Radiol. 2015;25(11): 3187–99.

37. Rajpurkar P, Irvin J, Ball RL, Zhu K, Yang B, Mehta H, et al. Deep learning for chest radiograph diagnosis: A retrospective comparison of the CheXNeXt algorithm to practicing radiologists. PLoS Med. 2018 Nov 20;15(11).

38. Topol E. Deep Medicine: How artificial intelligence can make healthcare human again. New York: Basic Books; 2019.

Recommended Reading

Pedro Domingos, (2015) "The Master Algorithm: How the quest for the ultimate learning machine will remake our world," (Basic Books, New York).

Stuart Russell and Peter Norvig, (2020) "Artificial Intelligence: A Modern Approach," 4th edition, (Prentice Hall, Upper Saddle River, NJ).

Terrence J. Sejnowski, (2018) "The Deep Learning Revolution," (MIT Press, Cambridge, MA).

Ian Goodfellow, Yoshua Bengio and Aaron Courville, (2016), "Deep learning," (MIT Press, Cambridge, MA).

Sutton and Barto, (2018), "Reinforcement Learning: An Introduction," (MIT Press, Cambridge, MA).

Gary Marcus and Ernest Davis, (2019) "Rebooting AI: Building Artificial Intelligence We Can Trust," (Pantheon Books, New York).

The Long Road to New Medicines

Nature is the healer of disease

Hippocrates, 460 BCE

Viewed through the lens of therapeutic discovery, humans have been on a 10,000-year quest to find natural or synthetic substances with medicinal properties. Much of this exploration, even within the context of modern pharmaceutical science, has been done by random experimentation with nature's *pharmacopeia*. Looking back over the millennia, the journey has been a never-ending search through the planet's immense catalog of some 350,000 plant species. The exploration thus far has yielded thousands of food substances, bioactive molecules, and vitamins that are beneficial for dietary health, and it has uncovered a much rarer set of plant-derived compounds possessing medicinal value, as proven by their ability to ameliorate symptoms or alter the course of illness and disease. Out of these, pain-relieving compounds derived from the opium poppy formed the foundation of the pharmaceutical industry two centuries ago.

Following the Renaissance and well into the nineteenth century, nearly all of the concoctions and herbal medicines that might have contained a pharmacologically active compound, such as morphine or codeine from opium, evaded purification by extraction and were presented in more soluble or bioavailable forms only through mixing with alcohol and other solvents. Plants produce a vast repertoire of chemical compounds, and the chemical entities responsible for delivering biological activity of medicinal value are small organic molecules. In Europe, advances in the rapidly developing field of chemistry aided pharmacists, as well as alchemists and charlatans, to obtain botanical extracts with greater purity. The new chemical processing techniques led to more consistent routes for the manufacture of pain-relieving medicines to patients and, unfortunately, to more potent drugs that could be distributed and abused by the general public. The successful extraction of morphine from opium latex was achieved by the German pharmacist Freidrich Surtürner in the early 1800s, and it was one of the early breakthroughs needed for the future industrialization of drug production. Emanuel Merck developed similar *alkaloid* extraction techniques, and he saw an opportunity to scale up the manufacture and marketing of opiate alkaloids for his family's company, starting in 1827. The move from "pharmacy to factory" took place during the 1830s in Germany, with the appearance of Merck's first factory on the eastern edge of Darmstadt, a town south of Frankfurt and situated near the Rhine River.

Pharmaceutical manufacturing roared to life within the German capitalistic environment as a result of the convergence of new chemical processing technologies and the Industrial Revolution. Isolation of single-ingredient drugs from botanical extracts became the method for obtaining pure medicinal compounds in Europe and then the United States, as alchemy and its toxic mixtures and elixirs slowly began to fade away in medical practice. The advancement of techniques in analytic and organic chemistry, together with the potential commercial value of natural extracts, led to a rush in isolation of other plant alkaloids, including caffeine and quinine (1820), nicotine (1828), atropine (1833), theobromine (1842), cocaine (1860), and muscarine (1870). Additional opiate alkaloids were also successfully separated out by Merck scientists and others (codeine, papaverine, and thebaine), cementing the poppy as the most important medical crop in both ancient and modern human history.

The journey of drug discovery entered a new phase in the late nineteenth century with the development of synthetic chemistry. The idea that new compounds, which do not naturally exist in nature, could be synthesized by altering substances of the natural world was relatively new. Physicians and medical practitioners in the West had only begun to abandon the Hippocratic concept that the four humors—blood, yellow bile, black bile, and phlegm—corresponded to four human temperaments: sanguine, choleric, melancholic, and phlegmatic. The alchemists had mixed together elements such as gold, mercury, and sulfur together with acids and an array of organic materials (dung, horsetails, and root vegetables to name a few) for their imaginative amalgamations. But they had not contemplated the creation of new structures outside of nature's preset catalog. Nor could they do so, as no adequate concepts or theory existed as to what constituted a chemical compound or the physical essence of life. Only with a scientific foundation built by Lavoisier, Priestley, Avogadro, and other chemists, together with Dalton's principles and the framework of atomic theory, put forth in 1802, could chemists envision the possibilities of tinkering with chemical structures present in nature. The design and testing of new compounds came about as the field of chemistry matured and industrial-scale manufacturing could produce an array of compounds for research and development purposes. Almost in lockstep, the experimental foundations of pharmacology were also being established in university settings. In 1905, John Newport Langley proposed the crucial concept of agonists and antagonists acting at a single receptor, adding to Paul Ehrlich's work on chemoreceptors.[1,2] Throughout the twentieth century, the *medicinal chemistry* expertise accruing in these industrial enterprises equipped drug makers with the tools to replace purified natural extracts with cheaper and more purpose-built synthetic molecules. The magic formula for successful drug discovery was the creation of an interdisciplinary research enterprise guided by clinical sciences, biochemistry, microbiology, and serendipitous findings in biology.

Biotechnology ushered in a third phase of discovery and rapidly became the driver of innovation for a new class of drugs known as *biologics*. With the tools of *recombinant DNA technology* built by molecular biologists in academic labs during the 1970s, a new generation of therapeutic development and manufacturing was enabled. The promise of biotechnology was the newfound power to synthesize protein products that were the precise imitations of those from human cells (see Chapter 4, "Gene Editing and the New Tools of Biotechnology"). Commercially important hormones, such as insulin and adrenocorticotropic hormone, which were derived from animal tissues, now had a new path to becoming therapeutics in a cellular factory. Starting with a protein's gene sequence information, cells could be engineered to generate these complex biological macromolecules. The pharmaceutical industry quickly scaled up production facilities to grow cells and purify these proteins as eventual biologic drugs. The race for the world's first biotechnology drug was won by Eli Lilly and Company in 1982, which worked with the biotechnology company Genentech and received FDA approval for their recombinant human insulin, trademarked as Humulin.

Monoclonal antibody technology was the other driver for the entrance of biologics into the therapeutic arena. The technology creates a novel cell type called a *hybridoma* that is utilized to produce antibodies derived from a single parent B cell of the mouse's immune system. The growth of genetically identical B cells (a clone) is driven by fusion with a tumor cell, which bestows immortality for continuous growth in culture. In 1986, the first mouse monoclonal antibody was approved by the FDA and marketed by Janssen as Orthoclone OKT3. The antibody-based drug was designed as an immunosuppressive agent, recognizing a T cell-specific protein known as CD3, and used to block T cell activation and combat acute rejection of organ transplants. Since OKT3's arrival, manufacturing processes for monoclonal antibodies have undergone continuous improvements via genetic engineering to remove mouse sequences and avoid potentially serious complications from unwanted immune reactions. The latest generation of therapeutic antibodies are produced using a hybridoma system from mice engineered to contain only human antibody genes, known as *fully humanized antibodies*. An alternative technology, known as *phage display*, allowed drug developers to discover and build antibodies in an entirely different way with genetic techniques and new cell lines.[3] The world's best-selling drug, Humira, was created by the latter method and is a fully human monoclonal antibody for rheumatoid arthritis.

Despite the complexities of biologic drug development, a steady stream of products has been introduced over the past several decades to treat cancer, autoimmune disorders, cardiovascular disease, and metabolic disorders. Biologics have come to dominate global drug sales; monoclonal antibodies and recombinant protein therapeutics together occupied eight of the top ten slots and generated nearly $75 billion in revenues in 2019. In the race to develop therapies for COVID-19, monoclonal antibodies were among the first drugs developed, which included Lilly's bamlanivimab and Regeneron's antibody cocktail REGN-COV2, used to treat President Trump in October 2020.

During this third wave of discovery, biotechnology companies became the engine of discovery. The major pharmaceutical companies (nearly all of those considered "Big Pharma" today) scrambled to reorganize and incorporate biotechnology expertise into their early discovery and pre-clinical development programs in an effort to complement their existing small molecule drug pipelines. By building on conceptual advances in biology and adopting the tools brought by molecular biology, genetics, and immunology, the drug hunters and medicinal chemists shifted their focus from exploring new chemical structures to the identification of new *molecular targets* as starting points. Exponential increases in biological knowledge and genomic data tied to physiological processes illuminated many potential sites of therapeutic intervention. These sites were in many cases novel drug targets tied to a biological activity that might have been discarded in prior decades as undruggable. To mine these potential treasures, chemists and biologists worked in concert and created a new drug discovery paradigm. The target discovery and validation steps were handled by biologists, who also created the high-throughput screening assays needed to search through vast chemical libraries to find compounds that could interfere with or modify the target's activity. Detecting drug candidates that had potency against the target in living cells is known as *hit discovery*. High throughput screening in biology became the foundation for hit discovery into the twenty-first century.

Biotechnology has risen as a shining star, not simply for pharmaceutical company profits, of which there have been plenty, but because of the myriad tools and versatility it brings to conquering disease. In the fourth phase of discovery, recent developments in cell and gene therapy, gene editing, cell reprogramming, and synthetic biology are on the cusp of providing the industry with surgical tools to precisely engineer therapeutics and cures, starting with our own cells and genes as the starting material. The main characteristic of the fourth phase is about the ability to reprogram cells, rewire the brain, rewrite genes, and redesign protein machines, all of which is really about the future of medicine.

FOUR ERAS OF THERAPEUTIC DISCOVERY

Therapeutic discovery has evolved through four eras.

Botanicals   Botanical compounds were found for medicinal, recreational, and religious use by random experimentation in nature's pharmacopeia.

Chemical Therapeutics   Chemical extraction and purification of botanical compounds were followed by novel compound synthesis and the birth of pharmaceutical manufacturing.

Biotherapeutics   Biotechnology companies introduced biologics manufacturing to produce an array of new therapeutic classes, including antibodies and recombinant proteins.

Therapeutic Engineering   Cell and gene therapies are engineered rather than discovered; technologies will emerge to produce designer biological molecules via gene editing; brain/behavior modification will occur with prescription software; precision delivery systems will enable new routes for treatment using viral vectors, nanoparticles, and nanorobots circulating in the bloodstream.

Medicine's Origins: The Role of Opium Since the Stone Age

Of all the remedies it has pleased almighty God to give man to relieve his suffering, none is so universal and so efficacious as opium.

Thomas Sydenham (1624–1689)

"Who is the man who can take his leave of the realms of opium?"

Charles Baudelaire in *Artificial Paradises* (1860)

The roots of medicine run deep in humankind's history, possibly coinciding with the discovery of opium. Archaeobotanical evidence indicates that the opium poppy was present in Neolithic cultures of Northern Europe, Spain, and throughout the Mediterranean. Opium's cultural significance is evident from Bronze Age figurines and jewelry adorned with poppy capsules to the billions of pills prescribed for patients in present-day healthcare systems. In various forms, opiate alkaloids have been constant human companions for medical, recreational, and religious use and of major economic value for thousands of years.

Compared to the Industrial or Information Revolutions of the nineteenth and twentieth centuries, the Neolithic Revolution is arguably the most important in human history—it was the gateway to civilization, bringing together technologies to plant and harvest crops, domesticate animals, and live in settlements. Innovation in medical practices and use of medicines to fend off the ravages of disease, famine, wars, and plague, plus the bevy of societal ills affecting physical and mental health, came extremely late in the game.

The rise of agricultural practices and their spread out of the Fertile Crescent by Neolithic agrarian settlers led to crop introductions across Western and Northern Europe more than 7,500 years ago. Analysis of carbonized plant material from several archeological sites in these regions establish the co-existence of many crops, such as various wheats, pea, lentil, barley, and flax that made their way along routes out of the Near East. The interesting exception is the opium poppy, which appears to have moved southeast over the Alps and back into the Mediterranean region much later. The dispersion of crops over Neolithic Europe is shown in Figure 3.1, which summarizes the routes that were taken into Europe, the crop species that were present, and zones where the earliest known poppy seeds were identified.

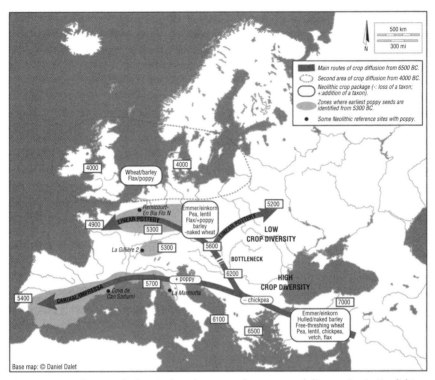

Figure 3.1: Early spread of agricultural crops and poppy seed discoveries in Neolithic Europe

With permission from [A. Salavert, agricultural dispersals in Mediterranean and temperate Europe. Oxford Research Encyclopedia of Environmental Science, 2017.[4]]

Whether cultivated or domesticated as a weed or plant crop, seeds from two *Papaver somniferum* subspecies are found at several linear pottery culture sites (the Linearbandceramik people) around 7,300 years ago.[5–7] One of the clearest indicators of the extent of the opium poppy's spread and its importance in Neolithic cultures was found in burial caves in the southern province of Córdoba in Spain.[8] Several bodies in burial chambers within the Cueva de los Murciélagos—a prominent prehistoric site known as the "bat cave"—were found buried together with small woven pouches containing poppy seeds.[9] The presence of poppy in several of these Neolithic locations implies that one or more of the cultures had domesticated a poppy species. Another fascinating discovery related to the spread and potential use of opium was uncovered at an archeological site in Raunds, 35 miles west of present-day Cambridge, England. A group of eight poppy seeds was found in a prehistoric garbage site, dated between 3,800 and 3,600 BCE, suggesting that opium might have been traded with people from the British Isles.[10]

Given the accumulating prehistoric evidence, the most enduring phase of therapeutic discovery thus appears to reach all the way back to the latter part of the Stone Age, when humans began agricultural practices and were also busy sampling plants and, in essence, searching for botanical compounds with medicinal value or for those that would provide pleasure. The latter encompasses hallucinogenic compounds that distort perception and were sought for religious and spiritual purposes, and all other manner of psychoactive compounds, including stimulants and depressants that alter the mind, mood, and behavior, many of which are the recreational drugs of today. Over time, this process led to the development of herbal medicine, and nearly all cultures have possessed a canon of herbal remedies. Of these, the most important in western medicine up until the seventeenth century was *De Materia Medica*, written by the Greek physician Pedanius Dioscorides between 50–70 CE. The five-volume work documented the use of over 600 medicinal plants and was the predecessor of all modern pharmacopoeias. The Materia was no doubt influenced by the botanical treatise of Aristotle's disciple and successor, Theophrastus, who constructed the first systematic description of numerous plants in his ten volume *Enquiry into Plants* (or *Historica Plantarum*), written between 350 and 280 BCE. The ninth volume recounted how to extract gums, resins, and juices for medicinal purposes. Aristotle and Hippocrates discussed the opium poppy's medicinal use, and Aristotle was aware of opium's hypnotic effects.

A number of pharmacopoeias from the Middle Ages were written by physicians practicing traditional Persian medicine. These included *The Canon of Medicine*, written by Avicenna in 1025, and works by Ibn Zuhr in the twelfth century and Ibn Baytar in the fourteenth century. In China, herbal medicine is thought to have had its origins nearly five thousand years ago. The earliest known Chinese pharmacopoeia, however, is *The Shennong Herbal*, a compilation that covers the use of 365 medicinal herbs dating back about 1,800 years ago.

In Western Europe, the mantel of authority in pharmacopeias shifted over to the *Dispensatorium* in the mid-sixteenth century, written by one of Europe's greatest botanists, Valerius Cordus (1515–1544), in Germany. Cordus was one of the first to begin systematic investigations into the medicinal plants that were described in *De Materia Medica*, publishing the results of this research in his *Historia Plantarum*, considered one of the most important of the "herbals" ever written. Cordus, in his own personal quest for botanical cures in Southern Europe, contracted malaria in Italy and died at the age of 29.

Herbal formulas and recommendations on the use of opium are found among the pages of nearly all of these pharmacopeias, dating back thousands of years to instructions on its use in ancient Egypt. The Ebers Papyrus,[a] from the sixteenth century BCE, contains a famous recipe on the use of opium as a way to put infants to sleep:

"Remedy to stop a crying child: Pods of poppy plant, fly dirt which is on the wall, make it into one, strain, and take it for four days, it acts at once."[11]

By the latter part of the Middle Ages, millennia after the Egyptians, Greeks, and Romans had incorporated opium into their medical practices, the use of the drug for medicinal and recreational purposes was widespread among the peoples of the Ottoman, Persian (Safavid), and British empires. Differences across cultures were emerging in use patterns by the sixteenth century, where people of the Ottoman and Persian societies, as well as the Mughal empire in India, preferred smoking or eating opium as the method of consumption. In Europe, Paracelsus (circa 1500) had invented laudanum, a concoction that mixed the opiate alkaloid substances with alcohol to make them soluble; this magic potion became the standard elixir for nearly two hundred years. The British physician Thomas Sydenham changed the course of opiate consumption with his version of laudanum, published in 1667, in which opium is mixed with sherry plus saffron, cinnamon, and cloves—a potent blend that became widely popular and was adopted across Europe and then later in America. In a rather striking coincidence of timing, Friedrich Jacob Merck purchased the Engel-Apotheke (Angel Pharmacy) in Darmstadt, Germany in 1668, the precursor to Merck KGaA, the world's first pharmaceutical company.[12] The doctors of the United Kingdom and Europe had formed a delicate alliance with the apothecaries, testing new formulations of unreliable, toxic, and questionable cures in a battle against the common maladies and plagues of the era.

The apothecaries played an increasingly major role in providing healthcare to the masses by the mid-eighteenth century in Britain, serving to prescribe and dispense medicines in addition to performing surgeries. Adam Smith, the famous Scottish economist, quipped in 1776 that the

---

[a] The document, found in a mummy's tomb, was sold to German Egyptologist Georg Moritz Ebers in 1873–74. It was thought to have been trafficked by American Egyptologist Edwin Smith, who purchased another famous Papyri, eponymously named.

apothecary was "the physician of the poor in all cases, and of the rich when the distress or danger is not very great."[13] One of the biggest challenges of the apothecaries was the delivery of more standardized dosages and purer drugs at a time when knowledge of chemical properties and extraction methods were not understood. Chemists had presented a competitive threat to apothecaries with their experimentation and adulteration of drugs for the marketplace. The apothecaries sought to protect their turf; physicians and others sought to reign in the apothecary's power in the United Kingdom by advocating for reforms and regulations via Parliament from the 1790s onward until the Apothecaries Act of 1815.[14]

Societal changes, global trade, scientific advances, and the evolving medical practice landscape at the turn of the nineteenth century in England, France, and Germany created a perfect climate to unravel the mysteries of opium. Friedrich Sertürner's pharmacological investigations into the sleep-inducing principle of opium led to the first successful isolation of the opiate alkaloid he named morphium, after Morpheus, the Greek god of dreams. The chemical name was later changed to morphine for consistency with the nomenclature of alkaline substances. The entire economic equation was about to change as the news of Sertürner's isolation of morphine as a crystallized substance was widely disseminated and available to the chemists and pharmacies of Europe. The relatively easy sourcing of opium from Turkey, Egypt, and elsewhere meant that the only bottleneck in reaping a fortune from drug production was in compound isolation. Emanuel Merck saw the opportunity and developed similar extraction protocols, which he published in 1826. This was followed by his company's plans to establish large-scale manufacturing of opiate alkaloids and others for commercial use, as outlined in his 1827 manual, *Pharmazeutisch-chemisches Novitäten-Cabinet* (Cabinet of Pharmaceutical and Chemical Innovations). Soon thereafter, Merck was producing both morphine and codeine as medicines for pain, cough suppression, and diarrhea. Figure 3.2 depicts the capsule of the opium poppy containing the "milk of paradise," the molecular structure of morphine, and the main alkaloid compounds found in the plant.

Raw opium and the opiate alkaloids were big business, and their growing economic value had enormous repercussions in the mid-nineteenth century. British sea merchants in the opium trade were selling an estimated 1,400 tons of opium to China through the port of Canton in the 1830s. Opium use was prevalent across all levels of society, and the Chinese government had a huge problem, with potentially 12 million people addicted by the end of the decade. After failing to negotiate a trade deal with the British, the Chinese shut down all opium importation, leading to the first Opium War with Britain in 1840.[6] In the United States, morphine had been available in a variety of forms since the American Revolution. A pharmacy in Philadelphia, Rosengarten & Co., was the first to establish the manufacture of crystallized morphine in 1832. The Civil War greatly accelerated the use of morphine in the United States, where an estimated 10 million opium pills and 2.8 million ounces of other opioid compounds were administered to wounded soldiers. Morphine could be self-administered with hypodermic syringes, which likely facilitated the high rates of opioid addiction among soldiers in the war and returning veterans. It is reasonable to assume that one of the main roots of the opiate addiction problem in the United States took hold on the bloody battlefields during the Civil War.

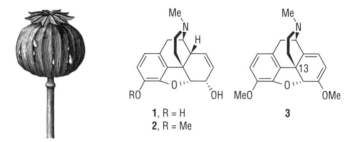

Figure 3.2: The opium capsule and opiate alkaloids

An illustration of the opium poppy capsule and the structures of morphine (1) with the substitutions of the parent compound at (R) that generate the main opiate alkaloids in the plant, codeine (2) where R equals hydrogen, and thebaine (3) shown to the right where R equals a methyl group.

Whether it was fortuitous or by destiny, painkillers, opioid drugs, and other narcotics were the pioneer medicines that laid the foundation for the pharmaceutical industry. Medicines were not available (from any rational scientific perspective) to target mechanisms of disease or infectious pathogens. Prior to the discovery and recognition of the role of microorganisms in infectious disease and the development of germ theory, the most important aims of medical practice were in helping a patient with quality of life and symptom relief. To that end, the demand from pharmacies, physicians, and the public were for more consistent doses of drugs with fewer side effects and greater potency, and the drug companies were poised to deliver. Merck, with morphine and codeine already in manufacture, added cocaine to its production line in 1862. At Friedrich Bayer & Co., management began research to find a chemical substitute for salicylic acid, a fever-reducing compound from the bark of the willow tree that produced unwanted side effects. Carl Deisburg, head of research at Bayer and Heinrich Dreser, head of experimental pharmacology, were also interested in creating a less addictive morphine or a codeine derivative. In 1897, Felix Hoffmann completed the first synthesis of acetylsalicylic acid, later marketed as aspirin. Arnold Eichengrün, who was the chemist in the group, wrote a letter in 1944 while in the Theresienstadt concentration camp that he had instructed Hoffmann to pursue acetylation of salicylic acid.[15] Eichengrün, a German Jew, received little credit during his lifetime for his role, which was buried to conform to Nazism. The acetylation of morphine led to diacetylmorphine, a more potent opioid than the parent molecule.[15,16] Although both of these substances had been described or patented previously, Bayer's aggressive marketing placed its brand on the most famous (aspirin) and infamous (heroin) drugs ever produced by the industry.

Industrial Manufacturing of Medicines

It would be an unsound fancy and self-contradictory to expect that things which have never yet been done can be done except by means which have never yet been tried.

Francis Bacon, *Novum Organum*, 1620

Time is the best appraiser of scientific work, and I am not unaware that an industrial discovery rarely produces all its fruit in the hands of its first inventor.

Louis Pasteur, 1879

The global pharmaceutical industry emerged from two very different entrepreneurial business models during the Industrial Revolution. The direct path came out of apothecaries or pharmacies that historically supplied medicines and had ambitions to manufacture more standardized versions of known drugs on a larger scale, such as what Merck in Germany and nearly all precursors to the big pharma companies did in the United States, as presented in Table 3.1. The second path was driven by synthetic chemistry, as coal tar dye companies in Switzerland and Germany sought to move into pharmaceuticals once the widespread industrial applications of chemistry were recognized. There are striking parallels between the emergence of the pharmaceutical industry during the Industrial Revolution and the rise of the technology sector 100 years later. In the same way that chemistry and hydrocarbon building blocks were the catalyst for pharmaceutical innovation, electrical engineering and integrated circuits were the drivers of advances in the computing era.

Table 3.1: Genesis of the Global Pharmaceutical Industry

COMPANY (ORIGIN)	YEAR FOUNDED	GLOBAL SALES, CONSOLIDATION STATUS (2019)	INITIAL PHARMACEUTICAL AND MEDICAL PRODUCTS
Chemical Companies			
Geigy (Swiss)	1858*	$48.6B, merged with Ciba; part of Novartis	Anti-inflammatories
Ciba (Swiss)	1859	Merged with Geigy; part of Novartis	Antiseptics
Hoechst (Germany)	1863	$40.4B; part of Sanofi	Antipyretic; antituberculin; novocaine; insulin
Kalle & Co.	1863	Acquired by Hoechst	Antifebrin
Bayer (Germany)	1863	$49.9B	Aspirin; heroin; barbiturates
Boehringer (Germany)	1885	$21.3B	Opiate alkaloids; lactic acid
Sandoz (Swiss)	1886	Merged with Ciba-Geigy; part of Novartis	Antipyrine
Bristol-Myers (US)	1887	$26.1B	Toothpaste; Sal Hepatica salts
E.R. Squibb (US)	1892*	Acquired by Bristol-Myers	Ether; Digitalis; curare
Pharmacies and Medical Supply Companies			
Merck KGaA (Germany)	1827*	$17.9B	Morphine, codeine
Beacham (England)	1842	$44.7B; Merged with SmithKline; part of GSK	Laxatives
Pfizer (US)	1849	$50.7B	Antiparasitics; Vitamin C
Schering (Germany)	1851	Acquired by Bayer	Atophan and Urotropin

Continues

Table 3.1 (*continued*)

COMPANY (ORIGIN)	YEAR FOUNDED	GLOBAL SALES, CONSOLIDATION STATUS (2019)	INITIAL PHARMACEUTICAL AND MEDICAL PRODUCTS
Wyeth & Co (US)	1860	Acquired by Pfizer	Glycerine suppositories
Sharp & Dohme (US)	1860 (1892)	Acquired by Merck (US)	Drug packaging; hypodermics
Parke, Davis & Co. (US)	1866	Acquired by Warner-Lambert, part of Pfizer	Tetanus and diphtheria antitoxins; cocaine
Eli Lilly & Co. (US)	1876	$23.1B	Medical kits; Insulin
Burroughs Wellcome (England)	1880	Merged with SmithKline Beacham; Part of GSK	Vaccines
Johnson & Johnson (US)	1885	$82.8B	Antiseptic bandages; contraceptives
Abbott (US)	1888	$33.3B#	Alkaloids; antiseptics
H.K. Mulford (US)	1889	Merged with Sharpe & Dohme; part of Merck	Smallpox vaccine; diphtheria antitoxin
Merck (US)	1891	$47.9B	Vitamin B12, cortisone, streptomycin
Takeda (Japan)	1895*	$29.7B	Antidiarrheals; quinine; saccharine
Roche (Swiss)	1896	$61.9B	Iodine; cough syrup; barbiturates

Sales figures are based on either Forbes list of the Global 2000: The World's Largest Public Companies (www.forbes.com/global2000/#f6c8ed7335d8, May 13th, 2020), company annual reports (Boehringer), or Dun & Bradstreet (Merck KGaA).

Abbott sales are those of AbbVie, formed when Abbott established AbbVie as separate pharmaceutical business in 2013.

* Companies with older historical roots prior to the introduction of manufacturing: Merck's original founding as a pharmacy was in 1668; Geigy was originally founded as a business in 1758; E.R. Squibb established a laboratory in Brooklyn in 1858; Takeda was founded in 1781 as a seller of traditional Japanese and Chinese medicines.

DEVELOPMENT OF THE PHARMACEUTICAL AND COMPUTER INDUSTRIES

The following were the confluence of events and economic conditions critical to the development of the pharmaceutical and computer industries:

Markets Demand pre-built in markets with enormous needs.

Capital Britain, Belgium, and Germany financed the growing industry.

Resources Raw materials for product manufacturing were plentiful and cheap.

Labor Growing workforce with technical training and skills.

Research and Development R&D activities embedded in successful enterprises to support continuous innovation.

Inventors and Pioneers Strong links between university and enterprises were a critical source of ideas and technical know-how.

The textiles industry was roaring by the mid-1850s, and Britain, Belgium, Germany, France, and Switzerland led the way, powered by mechanized manufacturing and steam engines. Natural dyes for fabric were expensive, in high demand, and only available from vegetable sources. Synthetic dyes would have an enormous, immediate market. Several companies along the Rhine River in Switzerland and Germany had been producing natural dyestuffs and were located in close proximity to textiles manufacturers. The formation of the German Customs Union in 1834 had broken down barriers between states in Germany and transformed the economy, allowing goods to flow freely. Investments into Germany from Britain and Belgium began to finance factories across the Rhineland.

It turns out that the pharmaceutical industry had its own mythical startup spaces like those in Silicon Valley in the early days of the personal computer and Internet business boom. Steve Jobs and Steve Wozniak worked out of a suburban garage to build a prototype of the first personal computer in Palo Alto, California, in the 1970s. Decades later, countless startups had begun out of apartments and small spaces across the San Francisco Bay Area in California. As the Industrial Revolution swung into gear in the mid-nineteenth century, primarily in the United States, Britain, Germany, and Switzerland, a new class of entrepreneur began explorations that led to the formation of two entirely new and related industries—chemical production and pharmaceutical manufacturing.

In the United States, the physician Edward Robinson Squibb started the Squibb Company in 1858, setting up a startup lab in his house in Brooklyn as he worked on ether production and his first business ideas around industrializing medically significant compounds. In 1856 in London, the teenage chemist William Henry Perkin set up his family's attic as a laboratory to conduct synthetic chemistry experiments in their east-end townhouse.

Of all of the elements needed for the ensuing firestorm of activity that surrounded the industrialization of drug making, it was the serendipitous discovery that Perkin made while attempting to synthesize quinine from chemical compounds in coal tar that provided the spark. Perkin was an eighteen-year-old chemistry student, and his mentor was August Wilhelm von Hofmann, a German expatriate who was by then an expert in the chemical composition of coal tar and organic synthesis techniques. Coal gas and solid coke were the main fuels used in homes and businesses across Europe, the United Kingdom, and the United States by the mid-1800s. Coal tar was the residual by-product of burning coal at high temperature to make coal gas and coke. It was therefore an abundant industrial waste. Hofmann correctly assumed that the hydrocarbon constituents of coal tar could be used in synthetic processes, and he had hypothesized that aniline, a chief component, could serve as a starting point in a synthetic pathway to quinine, the valuable and scarce anti-malarial compound from the bark of the South American cinchona tree.

Perkin's many attempts at producing quinine had failed. But on one occasion, he oxidized aniline in sulfuric acid with potassium dichromate and obtained the purple dye he named mauveine.[b] The first synthetic organic chemical dye had been produced, and both he and his mentor recognized the commercial value. However, the duo proceeded down different paths into history. To Hofmann's chagrin, Perkin left school and formed a small venture underwritten by his father, building a factory on the outskirts of London to produce the dye after he patented the process in the United Kingdom. Hofmann left the Royal College of Chemistry where he was the founding director (now known as Imperial College London) and returned to Germany, where he co-founded the German Chemical Society and was a professor at the University of Berlin. It would be Hofmann's return that gave Germany the decided edge in expertise and training as the new coal tar dye and chemical companies sprouted up along the Rhine River.

---

[b] The chemical reaction products in coal tar were likely a mixture of aniline, o-toluidine, and p-toluidine, and his resulting product was not a single compound but a mixture of compounds with a common polycyclic chromophore—the structure responsible for the brilliant mauve color.[17]

Germany came to dominate global chemical and pharmaceutical manufacturing up until World War II. How was this accomplished? The primary factor was a strong scientific foundation and a focus on intensive research.[18,19] Freidrich Bayer & Co. was the exemplar for the successful business model. In 1863, the founders began by performing experiments in their kitchens, attempting to create aniline dyes based on the known art of the time. However, the first-generation coal tar dyes lacked the properties needed for textiles—colorfastness as well as light-fastness. Solving any of these problems required research and staying informed of the latest developments in the scientific community. One of the stumbling blocks was that the chemical structure of molecules, such as benzene, was unknown. They had knowledge of the molecule's empirical structure, C_6H_6, but how were carbon-carbon bonds distributed in the molecule? One of the giants in organic chemistry—the German-born August Kekulé—unlocked the major secret of organic chemical structures with his 1865 publication of the structure of benzene (a six-carbon ring compound with alternating double bonds, which was the backbone of aniline, both pictured in Figure 3.3).[20] The applied chemists at Bayer were thus able to use the information to construct, or *derivatize* in chemist's parlance, an array of dyes off of the simple *aromatic* and *aliphatic* compounds present in the coal tar mixture.

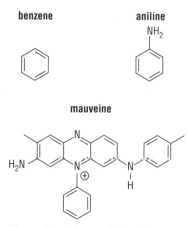

Figure 3.3: Chemical foundations of the coal tar dye industry

Chemical structures of benzene, aniline, and mauveine. Kekulé's structure and theory of benzene, Hofmann's building block for dyes, and Perkin's synthesis of mauveine were the scientific inspirations that launched the coal tar dye industry.

Once the aniline-based dye making process was established, another market was created for the dye-making chemicals isolated out of coal tar, since not all dye manufacturers had industrial-scale processing expertise.

Many companies were founded during that time period to capitalize on the new field of organic chemistry and vast markets. Along the way, only innovative, research-oriented companies survived. In retrospect, it is clear that fierce competition for trademarks forced the dye businesses to seek opportunities in international markets; patents could be skirted or ignored, as happened in Switzerland, and Germany had not yet put patent protection into place. Perkin's patent was practically useless, as he sought protection only in the United Kingdom. Pharmaceutical companies later emerged as divisions of these coal tar dye chemical companies. The organic compounds—the aliphatic and aromatic building blocks—were to be used extensively and profitably in medicinal chemistry.

Paul Ehrlich and the Birth of Chemotherapeutic Drug Discovery

For those late-nineteenth century chemical companies and the dyestuff industry to move from chemicals and fabric dyes to human medications, they needed clinical science expertise from medical disciplines whose focus was on the links between chemistry and biology. Two fields arising at nearly the same time that could provide this bridge were pharmacology and microbiology. The undertaking of pharmaceutical research and development required an interdisciplinary approach, and academic researchers at the forefront of drug discovery at the time, most prominently Paul Ehrlich, understood this intuitively. Creating this type of atmosphere in industry was only beginning to form in companies such as Bayer, Merck, Hoechst in Germany, and literally nowhere else in the world.

Ehrlich's contributions to medicine were immense; he led research into groundbreaking therapeutic discoveries and laid down several of the foundational theories of immunology and pharmacology. He won a share of the Nobel Prize for Physiology or Medicine in 1908 for "outlining the principles of selective toxicity and for showing preferential eradication of cells by chemicals."[21] As the originator of the concept of chemotherapy, Ehrlich had the idea to create a "zauberkugel," a chemical magic bullet that could selectively target and destroy pathogenic organisms without harming host cells, much like a precision-guided weapon. Ehrlich saw promise in aniline dyes, such as methylene blue, with which he had great familiarity dating back to his graduate work in histology and those of his cousin, Karl Weigert, who pioneered their use for studying bacteria. Since methylene blue could stain malaria-causing plasmodia and therefore bind to these microorganisms, he reasoned that dyes might be a suitable therapeutic device, one that could interact with a parasite-specific receptor. Ehrlich obtained a quantity of the dye from Hoechst, the chemical company with which he established a long-term collaboration. Lacking any adequate animal model, he arranged to administer the dye to two patients, both of whom were inmates of the Moabit prison in Berlin and suffering from malaria.[c] The two men recovered, and the plasmodial parasites disappeared from their blood. This was an astonishing feat, and Ehrlich's small clinical test in 1891 became the first use of a synthetic "drug candidate" to target an infectious disease agent in humans.[22]

The malarial drug studies came to an end as Ehrlich was still in the middle of another landmark clinical science effort under the direction of Robert Koch, the physician and bacteriologist responsible for developing the germ theory of disease. Koch wanted Ehrlich to team up with Emil Behring, who had been building on his groundbreaking work using "serum therapy" for diphtheria. An international race was underway to produce an effective serum against the deadly diphtheria toxin out of immunized animals, and Koch had tasked Ehrlich to work on standardizing the preparation. Clinical tests with these immune sera in 1894 were successful, and Hoechst had arranged for the commercial rights to develop supplies and market the "diphtheria remedy synthesized by Behring-Ehrlich" throughout Europe. In the United States, Parke-Davis was the first to begin manufacturing the immune serum. Behring received the first ever Nobel Prize in Physiology or Medicine in 1901 for the remarkable clinical results achieved against the dreaded childhood disease and the development of the theory of humoral immunity.

Ehrlich had been nominated but not selected for the very first Nobel prize, and he was also short-changed in the royalty agreement made with Hoechst. But Ehrlich had moved on. He was fortified by his initial success against the malarial parasite, and his research into dye-based drugs against other pathogens continued unabated for the next two decades in the laboratories and research institutes he directed in Frankfurt. By the turn of the twentieth century, Ehrlich had a full-blown operation going to discover new antimicrobial drugs. With a group comprised of organic chemists, microbiologists, and pharmacologists, he managed what were

---

[c] Ehrlich's early histological work with methylene blue demonstrated that the axons of living nerves could be selectively stained by the dye. The inmates he treated at Moabit prison also had neurological conditions causing pain, and Ehrlich was interested in testing the effect of the dye compound on pain reduction in addition to its anti-malarial properties. It was reported to have worked as both an analgesic and antiparasitic drug. The use of methylene blue never took hold as a malarial medicine, however, as quinine was comparably much more effective.

essentially the first drug screening programs and an operational model adopted by the emerging pharmaceutical industry.

Trypanosomes were the next parasitic target of Ehrlich's discovery operation. A devastating sleeping sickness epidemic had swept across Africa between 1896 and 1906, resulting in somewhere between 300,000 to 500,000 deaths. Colonial powers, including Britain, Germany, and France, were eager to combat the widespread illness and the danger it posed to their continued colonialization efforts. Work on trypanosomes by Alphonse Laveran at the Pasteur Institute in Paris demonstrated that these organisms could be easily propagated in laboratory animals. Ehrlich and his Japanese colleague Kiyoshi Shiga established mouse colonies and tested hundreds of new compounds against different trypanosome species, and one dye, Nagana Red, had shown promising activity in infected mice. Ehrlich sought dyes with better drug properties but with structures similar to Nagana Red from another one of the dye companies, and through additional medicinal chemistry and drug screening efforts wound up discovering Trypan Red. Testing these benzopurpurine-based dye compounds in a series of experiments revealed a new facet that would come to forever plague the pursuit of antibiotics: the microorganism had developed drug resistance and the mice were no longer spared of disease by treatment. Veterinary field tests in Africa confirmed these laboratory observations.

Ehrlich's most enduring achievement in chemotherapeutic drug development came in the discovery of the first drug to treat syphilis, a dreaded affliction that rampaged periodically across the globe. The compound created in his laboratory became the first semi-synthetic antimicrobial drug in history. The causative agent of syphilis had been identified by Eric Hoffmann and Fritz Schaudinn, two of Ehrlich's Berlin colleagues, in 1905. They determined that the parasitic pathogen was a spirochete, a bacterium known as *Treponema pallidum*. Hoffmann suggested that Ehrlich use arsenical compounds to treat syphilis patients, as these spiral-shaped bacterial pathogens shared biological characteristics with trypanosomes. From what we know today, the suggestion was ludicrous, as these are completely dissimilar organisms, aside from their shared parasitic lifestyles.

One of the first therapeutic approaches that had some efficacy against trypanosomes was arsenic acid, a derivative of the metallic poison that had been synthesized in 1859 by the French biologist Pierre Jaques Antoine Béchamp. An extensive medicinal chemistry effort was launched by Ehrlich and his colleague Alfred Bertheim to synthesize less toxic but therapeutically active arsenical compounds against syphilis. Ehrlich's working theory was to combine two components together, one that bound to the organism and another that caused the organism's eventual destruction—in this case, the arsenic poison. Hundreds of molecular "lead" candidates were created for the next stage of screening, which took years.

In drug discovery, one of the bottlenecks to advancing drug candidates involves the use of *in vivo* pharmacological studies, where dosing regimens, evaluation of pharmacological properties, and toxicity are analyzed in laboratory animals. The library of new compounds Ehrlich and Bertheim created was tested in rabbits by Sahachiro Hata, the Japanese student who had just arrived in the lab. After a series of screening campaigns during which Hata tested hundreds of novel chemical structures, he finally identified arsphenamine—compound 606—as having the best activity against the infection in rabbits. Clinical trials were started to test the potential therapy in syphilitic patients, and the duo reported their first results with arsphenamine in 1910. The stunning announcement led to enormous demand from clinicians, with Ehrlich's institute sending out 65,000 free samples so that further clinical trials could be done. Soon thereafter, Hoechst landed a marketing agreement for the drug under the name Salvarsan with the clever tag line "the arsenic that saves." Ehrlich's dream of a magic bullet had come to fruition with worldwide acclaim.

Ehrlich's theories and methodological approaches were the entry gates to the modern practices found in drug discovery and development. At the beginning of the twentieth century, his receptor theory, chemotherapeutic principles, and ideas about mechanisms of drug action were at once revolutionary and controversial. Rivals at the Pasteur Institute were skeptical of receptor theory. Scientists of the day had many competing ideas about how drugs worked, which are now understood as entirely wrong, but these remained in circulation for decades.[16] Ehrlich had crystallized the concept that a drug could bind to a receptor (or some other target) and exert an action on cellular physiology.

Throughout the early development of the pharmaceutical industry and therapeutic discovery, the clinical research of Koch, Behring, and Ehrlich in Germany and Laveran, Mechnikov, and others at the Pasteur Institute in France exemplified the powerful turn toward a scientific approach to medicine. The scientific process involves an interaction between theory and observation; applied to medicine, it is the interplay of treatment and clinical outcomes. Ehrlich conducted his drug discovery research in a framework where a clinical hypothesis could be tested first in animal model systems that could provide proof of success or failure, with endpoints that were clear and predefined. This early discovery stage has also been defined by a strong trial and error component, which has remained largely unchanged over the past century. In Ehrlich's work on sleeping sickness and syphilis, the pathogens were known, and his idea was to selectively target and destroy the microorganisms with a magic bullet. He had no idea as to the nature of the molecular target with which the compounds would interact, or of any knowledge of biochemical processes at play that could lead to their destruction. By trial and error testing of thousands of compounds, at various doses, he was eventually able to stumble upon one or two that produced the desired outcome. After Salvarsan, it would be decades before scientists made further serendipitous discoveries to uncover the antibiotic properties of sulfa drugs and then rediscover penicillin.

The Pharmaceutical Industry: Drugs and War— New Medicines in the Twentieth Century

> *The abundance of substances of which animals and plants are composed, the remarkable processes whereby they are formed and then broken down again have claimed the attention of mankind of old, and hence from the early days they also persistently captivated the interest of chemists.*
>
> **Emil Fischer, Nobel Lecture, 1902**

The German chemical industry, including companies such as Bayer and Hoechst with pharmaceutical divisions, pursued global commercial dominance through cartel agreements, patent protection, and heavy investments in research and manufacturing technology during the first half of the twentieth century. In some respects, this was to be the blueprint for the big pharmaceutical companies that emerged from this era and still predominate today (see Table 3.1). Carl Duisberg, who had championed the research investment into early synthetic medicines at Bayer, was also a central figure in the cartelization of the German dye industry from pre-World War I into the Nazi years. The powerful monopoly control that trusts in the United States, especially the Rockefeller Standard Oil Trust, had exerted over competitors and on pricing, production, and distribution inspired Duisberg to align chemical companies to a similar model in Germany. Bayer, together with Agfa and BASF, formed the Dreibund cartel in 1905–06. Three other German chemical companies, which included Hoechst, had followed suit to form the Dreierverband cartel. After World War I, Duisberg and Carl Bosch, head of BASF, urged the formation of the notorious I.G. Farben, created out of Germany's big six chemical companies in 1925. From that point forward until World War II, I.G. Farben was the biggest enterprise in all of Europe and the fourth largest in the world, behind General Motors, United States Steel, and Standard Oil of New Jersey.

The practices of the German chemical cartels, and especially of I.G. Farben, were to have ghastly and catastrophic consequences for the world. Germany used advances in the chemistry labs of the cartel members to overcome its dependence on foreign supplies, such as for Chilean nitrates. Its chemists achieved a key milestone with the development of the Haber-Bosch process for fixing nitrogen, which enabled the mass production of agricultural fertilizers and high explosives. The chemical companies also produced poison gases used in war: mustard gas, phosgene, and chlorine during World War I and Zyklon B, the cyanide-based gas used by the Nazis to commit genocide against millions of Jewish people in concentration camps during World War II. I.G. Farben was a primary contractor for Hitler's war machine, and the company used slave labor at the I.G. Auschwitz concentration camp for production of synthetic fuels, plastics, and other materials. After World War I, chemists, engineers, and managers of the various cartel companies played an increasingly important role in Germany's nationalistic ambitions and its move toward *Wehrwirtschaft*, a militarized economy.[23]

Cartel practices and the economic power they exerted had a profound impact on world geopolitics, but they also propelled dramatic changes across the medical and pharmaceutical industries, particularly in the United States and Britain. Prior to World War I, very little drug discovery research had been conducted in these countries. Instead, industrial chemists were hired by companies to evaluate the purity of medicines coming under regulatory statutes of the 1906 Food and Drug Act in the United States. The need for synthetic chemists and research was ultimately driven by war and the Allies' vulnerability to Germany's cartels. The start of World War I led to severe medical supply shortages on the battlefield, as German exports of surgical equipment (accounting for 80 percent of supply in the United States) and production of drugs including Salvarsan, Aspirin, Luminal, and Veronal (sedatives), and Novocaine (an anesthetic) were limited or altogether halted. As new weapons and the unique conditions of trench warfare produced devastating casualties, a full-blown medical emergency raged on.

Getting drugs, antiseptics, anesthetics, and surgical supplies to the battlefields in Europe was of paramount importance. Four thousand miles of trenches, from the English Channel to Switzerland, created logistical challenges and medical nightmares. Antibiotics were still a generation away. Wounds of any sort led to lethal infections in the watery trenches due to contamination with *Clostridium perfringens*. Britain used coal tar chemistry to make antiseptics, and sodium hypochlorite was discovered to be effective against the deadly bacterium. German soldiers had access to drugs for sedation and analgesia produced by Bayer, Merck, and Hoechst—in particular the powerful barbiturate Veronal, Aspirin, antipyrine, phenacetin, and pyramidone. Other sedating medications were valerian, bromide salts, chloral hydrate, and paraldehyde. During the latter stages of the war, all sides experienced severe shortages of morphine and heroin. Infectious disease accounted for 50 percent of deaths from the war, with a myriad of causes: trench fever (from bacteria transmitted by lice), typhoid (from rats), measles, influenza, meningitis, dysentery, as well as venereal diseases, namely, syphilis and gonorrhea.

Salvarsan was the most widely prescribed drug in the world on the eve of the war; its production was soon shut down at the Hoechst plant in Germany. In England, the United States, Canada, France, and Japan, chemists worked under time pressure to develop methods for synthesizing arsphenamine (Salvarsan), with great success. Jay Frank Schamberg and his colleagues at the Dermatological Research Laboratories in Philadelphia had produced their version of arsphenamine (Arsenobenzol) by June 1915. After obtaining a license from the Federal Trade Commission to bypass the German patent in 1917, they became a chief supplier to the Army and Navy of the syphilis drug.[24] British arsphenamine (Kharsivan) was produced by Burroughs Wellcome, apparently within weeks of the outbreak of the war.[25] These wartime emergencies spurred government and industry to work together, bypassing cartel protections, establishing more standardized manufacturing procedures, and creating regulatory approaches for clinical trial evaluation of medicines under development.

World War I also changed the course of the pharmaceutical industry after foreign assets were seized in the United States as a result of the Trading with the Enemy Act 1917. United States subsidiaries of German companies became owned by new American business interests or put up for sale, including Merck, Bayer, and Hoechst, as well as thousands of German chemical patents and trademarks (including Bayer's Aspirin). Similar measures were undertaken as Great Britain entered the war. German medical trademarks were suspended, and the British government sought brainpower and chemistry expertise from 40 different universities and technical colleges. Professors and students labored to uncover active ingredients of important medicines and team with the nascent pharmaceutical industry to manufacture drugs.

Following the war, the German cartels moved aggressively to reform and reassert control over the chemical industry, in part for nation-building and also for regaining economic power. The chemical firms as a trust were reported to have raised one billion marks in capital to build plants and increase the output of nitrate.[26] As a result of the Treaty of Versailles, Germany owed 20 billion gold marks (5 billion dollars at the time) payable by May 1, 1921, as an installment of war reparations to the Allies, with bonds issued later for tens of billions more with interest.[27] Germany was to depend upon the industrial and economic strength of the chemical companies during its rebuilding phase.

Another consequence of the war and terms of the peace treaty was that Germany was forced to give up its four African colonies, which were Togo, Cameroon, German Southwest Africa (present-day Namibia), and German East Africa (Rwanda, Burundi, and Tanzania). In a somewhat bizarre twist, the cartel attempted to use its medical trade secrets to seek the return of the colonies back to Germany. It had been reported in the *New York Times* (1922) and later in the *British Medical Journal* (1924)[28] that Bayer scientists and representatives of the German government had sought to leverage their discovery of a cure for sleeping sickness, African trypanomiasis, for geopolitical gain.

Well before the war broke out, Bayer had recognized the potential political importance of anti-parasitic drugs in colonial Africa. Bayer's chemists had picked up where Ehrlich's earlier work left off on trypanosomes and Trypan Red around 1904. Following more than a decade of work, Bayer scientists had developed and screened more than 1,000 dye-based drugs with increasingly complex structural formulas, and they finally obtained the most active one, named suramin, sometime during 1916–1917. The shorthand name for the compound was Bayer 205, which was eventually marketed under the name Germanin. The medicinal triumph against the parasite meant that British and French colonial activities and occupation of territories where the tsetse fly (the vector for the parasite) was found could proceed without fear of disease. To no one's surprise, the British government rejected the German's insidious attempt at extortion. Scientists in France and England had already begun work to uncover the structural formula of Bayer 205, which was ultimately published by Ernest Fourneau from the Pasteur Institute in 1924. The Nazis later made a propaganda film during World War II entitled *Germanin: The Story of a Colonial Deed*, unironically depicting a German hero and the country's humanitarian efforts in Africa. Remarkably, the 100-year-old drug remains on the World Health Organization's list of essential medicines.

From Synthetic Antibiotics to the Search for New Drugs from the Microbial World

The discovery of microbes as the causative agents of infectious disease led to many of the initial breakthroughs in therapeutics and provided biologists, chemists, and the pharmaceutical companies with financial incentives to innovate and conquer these human maladies. Prior to Louis Pasteur's debunking of spontaneous generation, Robert Koch's research on the bacterial origins of disease and the epidemiological observations made by John Snow on the cholera epidemic, there was no commonly understood linkage of microorganisms to disease. The mainstream view was that "bad air" somehow brought about sickness—the miasma theory—that laid blame to foul odors emanating from corpses, hospitals, and other organic material. How the human body worked and how medicines acted, or what led to disease, was entirely unknown. Physicians as well as dye company chemists were entirely in the dark up until 1882, when Koch determined the causal linkage between a bacterium and tuberculosis. In rapid succession, the identification of at least 20 infectious agents underlying diseases such as cholera, typhoid, anthrax, and syphilis followed in what was to become known as the golden age of bacteriology. Despite the early successes of Ehrlich and the Bayer scientists with targeted antiparasitic drugs, a great need existed for a truly "broad-spectrum" antibiotic that could destroy any class of bacteria.

The quest for new antimicrobial drugs was hotly pursued in Germany and enhanced by the unique relationship that academic researchers had with the chemical companies, the most important of which had banded together as I.G. Farben in 1925. A pathologist at the University of Münster, Gerhard Domagk, had taken a temporary leave from his University position to work at a research institute set up by I.G. Farben to identify promising aniline-based drugs within the field of bacteriology. Domagk's first success was in 1932 when he discovered an antibacterial compound later sold by Bayer AG as the skin disinfectant Zephirol. His second project was groundbreaking. Years of drug screening experiments led down a path to a major new drug class—compounds that could effectively destroy streptococcal, staphylococcal, and gonorrheal bacteria—the first broad-spectrum antibiotic. The first compound was sulfamidochrysoidine, a potent red aniline dye that killed streptococcal infections in animals and humans. Domagk treated his daughter's streptococcal infection for his first clinical test. After two years of formal clinical testing, the drug was placed on the market in 1935 and called Prontosil. Domagk was awarded the Nobel Prize in 1939 for the discovery. The Nazi regime forced him to decline acceptance at the time, but he later received the diploma and medal after World War II.

An important mystery remained to be solved concerning Prontosil's action, since bacteriological tests in the laboratory failed to show any antibiotic properties of Prontosil. In France, researchers at the Pasteur Institute unexpectedly discovered the reason why. The active substance was actually a metabolite of the parent compound, the well-known and simpler molecule sulfanilamide. This surprise finding opened the exploration of so-called sulfa drugs and highlighted the importance of *pharmacokinetic* investigations that are central to all small molecule drug discovery operations today. Over the ensuing decades, sulfanilamide became an important structural skeleton onto which new therapeutics were built, including antibiotics, anti-hypertensives, diuretics, and anti-diabetics.[18]

Soon after Prontosil's introduction, the rediscovery of penicillin opened another major era in antibiotic drug discovery—compounds derived from the earth's microbes. Of course, the microbial drug era had begun back in 1928 when Alexander Fleming realized that a compound produced by a contaminating mold inhibited the growth of his *Staphylococcus aureus* cultures. The active compound, produced by *Penicillium notatum*, was named penicillin, and the result was quietly published in 1929.[29] Ten years later, Florey and Chain's team at the University of Oxford pursued Fleming's finding and completed their definitive work on penicillin in 1941, and the two men together with Fleming shared the Nobel Prize in Physiology or Medicine in 1945.[30,31] Penicillin was an incredibly powerful medicine, but it was only produced in vanishingly small quantities by the strain used by Fleming. The desperate conditions of the war in England necessitated the Allies to work together to figure out industrial-scale production of the drug. Florey and his colleague Norman Heatley traveled to the United States in 1941 and persuaded the United States Department of Agriculture and industry laboratories to produce penicillin in a collaborative effort throughout the war years. Strain engineering and growth modifications improved yields by 100-fold in two years. In the United States, Pfizer, Merck, Squibb, and Winthrop Chemical were key manufacturers, and the list grew to nine companies with funding from the War Production Board in 1943.

Many of these same pharmaceutical companies established in-house microbiology facilities and research divisions to find new microbial compounds. Soon after penicillin, other antibiotics such as streptomycin (1943), chloramphenicol (1947), tetracycline (1948), and erythromycin

(1952) were isolated from different *Streptomyces* species. University and industry laboratories continued their work on developing new mutant fungal strains for penicillin, switching to *Penicillium chrysogenum* in 1951, and eventually created the super strain Q-176, the mother of all future penicillin production strains. Microbes have provided the starting points for most of the major classes of antibiotics, including the β-lactams (penicillins), aminoglycosides (streptomycin, neomycin), macrolides (azithromycin, erythromycin), tetracyclines, rifamycins, glycopeptides, streptogramins, and lipopeptides. Since 2000, there have been 20 new antibiotics launched to treat infections in humans: only five of these represented new compound classes of synthetic or semi-synthetic origin. The bacterial defense mechanisms that evolved over a billion years to produce these compounds against their environmental competitors are nearly perfect weapons against other bacteria. Chapter 4 will introduce another bacterial defense system employed against bacterial viruses that has emerged as one of the most promising tools for human gene editing-based approaches for targeting disease: CRISPR.

It is nearly impossible to overstate the impact that penicillin's discovery has made on the pharmaceutical industry, medical practice, and global health. The drug companies finally had a new route to therapeutics, mining the vast repository of compounds produced by microbes. The biochemical properties of these early drugs as inhibitors of protein and DNA synthesis came into clear view with the discovery of DNA and the mechanisms of DNA and protein synthesis. The early research provided a direct link between the drug's interactions at the molecular level and its therapeutic outcomes. The inadvertent consequence of the low yield of penicillin from Fleming's strain led to microbial genetic engineering and the first attempts at re-engineering metabolic pathways to produce compounds from organisms instead of in test tubes. For medicine, this antibiotic wellspring meant that ordinary physicians had at their disposal potent weapons to cure patients from the most common to the rarest of bacterial infections. Today, approximately 25 percent of the WHO model list of essential medicines are small molecule drugs that treat infectious disease. However, the never-ending battle with superbugs like MRSA and antimicrobial drug resistance has again placed antibiotic drug dis-covery back on the long list of urgent healthcare system needs. From a global human health perspective, the introduction of penicillin and its heirs from World War II onward has saved the lives of countless millions. Prior to antibiotics, mortality reduction was brought about principally by improvements in public health measures and personal hygiene, better nutrition, and raised living standards. Clinical medicine's use of antibiotics has reduced the impact of former plagues and provided important insights on how to tackle humanity's other scourge emerging from within: cancer.

Developing Therapeutics for Cancer

The search for genetic damage in cancer cells and the explication of how that damage affects the biochemical function of genes have become our best hope to understand and thus to thwart the ravages of cancer.

J. Michael Bishop, Nobel Lecture 1989

The development of the pharmaceutical industry in the half-century spanning the introduction of Bayer's aspirin and heroin out to the mass production of Fleming's penicillin proceeded largely in a vacuum of biological information, especially on the causes of human disease. For combating infectious disease agents, the industry's synthetic chemists and their academic collaborators tested compounds in a trial-and-error fashion on infected laboratory animals and then in humans without knowledge of the candidate drug's mechanism of action and without any way of predicting its safety or its therapeutic efficacy. A similar course of action was initially taken with anti-cancer drugs. Ehrlich's ideas on chemotherapy and selective targeting guided the emerging field in the early 1940s and 1950s toward cytotoxic drugs. There was precedence and plenty of anecdotal evidence to suggest that destroying distinct cell types with a drug could work. But could chemical compounds possess any selectivity that would not result in damage to the rest of the cells in the body? The first clue came in the most awful of circumstances, as a result of the use of chemical weapons by the Germans at Ypres on July 12, 1917. The release of sulfa mustard gas rained down onto the battlefield and killed 2,000 soldiers, leaving surviving victims with horrific burns and damaged gastrointestinal tracts, while also completely obliterating cells in the bone marrow. Two American pathologists later evaluated the survivors of the gas attacks and documented the destruction of lymphoid and blood tissues, along with the bone marrow, the site of origin of white blood cells. It was unclear why these cells were so vulnerable. The results were published and then forgotten.[32]

The clinical observation on the selective cellular effects of a chemical toxin was the first clue that targeted destruction of a cell population was achievable. Decades later, the United States government, fearful of new chemical weapons that might be put to use, began a top-secret program in 1942 to characterize these agents. Louis Goodman and Alfred Gilman at Yale University took on a top-secret research project to study the biological effects of sulfur and nitrogen mustard compounds. After careful studies in mice and rabbits, the first patient for nitrogen mustard therapy was selected, a 48-year-old man with lymphosarcoma. The protocol called for 10 consecutive rounds of "chemo." After the second round, the various tumors had visibly shrunk; by the tenth, the tumors had disappeared. The very first chemotherapy success had been achieved for lymphoma. Gilman and Frederick Philips wrote about the underlying cellular actions of chemotherapy in *Science* in 1946, after the first clinical report was released publicly:

> *Furthermore, cellular susceptibility to these compounds appeared to be related in a general way to the degree of proliferative activity. With the conviction that only with an understanding of the basic mechanisms of cellular action could significant advances be made in the treatment of vesicant war-gas casualties, the study of the actions of the sulfur and nitrogen mustards on fundamental cell processes was pursued. . .Cautious preliminary trials have also been made of the possible value of the nitrogen mustards in the treatment of neoplasms, in particular those of lymphoid tissues.*
>
> Alfred Gilman and Frederick S. Philips, *Science* April 5, 1946

The nitrogen mustards act to cause extensive damage to DNA by alkylation and interstrand cross-linking of DNA; this chemical modification is detected by genomic surveillance mechanisms that then trigger programmed cell death.[d] Drug screening programs and clinical trials continued for decades in attempts to identify new compounds of this class. However, the most commonly used DNA alkylating agents in cancer therapeutics today are still the nitrogen mustards, which include cisplatin, cyclophosphamide, and mechlorethamine, the original compound used in the early clinical trials of chemotherapy.

Antifolates and the Emergence of DNA Synthesis Inhibitors

The launch point for a rational chemotherapeutic approach to destroy cancer cells of the human body, instead of foreign bacteria, would only follow after a deeper understanding of cell biology and principles of pharmacology. This would require two things. First, researchers had to obtain essential knowledge about cellular biochemistry and mechanisms

---

[d] The cellular response to DNA alkylating agents is quite complex. Depending on cell type, DNA alkylating agent and cellular state, responses may involve autophagy, cell division blockade, and induction of DNA repair to promote tumor cell survival. See Bordin et al.[33] for a detailed review.

of growth control. What are the molecular targets for inhibiting cancer growth? The second, no less critical piece, was the feedback from clinical observers on drug responses, dosing schedules, toxicity, and unexpected clinical findings. These two components of drug development are enshrined in the modern drug research paradigm: the target discovery and validation activities found within preclinical development and the clinical studies and phased clinical trials overseen by clinical development. The pharmaceutical companies had not ventured far into cancer therapies and largely left innovation to academic and medical institutions. The ideal innovator would be a physician-scientist that could maneuver intuitively between the two domains, and that person was Sidney Farber, a pediatric specialist working on childhood leukemias while at Harvard Medical School and The Children's Medical Center in Boston.

Hematologic cancers result from mutations in DNA leading to uncontrolled proliferation of white blood cells. In lymphoma, the cells reside in lymphoid tissue. Leukemias arise in bone marrow through distinctive sets of mutations in precursor white blood cells. Farber proposed a completely different line of attack for this type of cancer in the early 1940s, aimed at a metabolic pathway. There was evidence that a vitamin deficiency might be linked to acute leukemias as folate deficiency was noted in the serum of patients with fast growing tumors. Farber injected folic acid-like compounds (teropterin and diopterin) into a few leukemic children, and to his dismay he saw that the treatment "accelerated" the leukemia. Witnessing that result, he pivoted to testing antagonists of folate. He received several compounds that had been synthesized by scientists at Lederle Labs in New York and found them to be growth inhibitors. Farber decided to test the best one, aminopterin, in a small clinical trial for acute lymphoblastic leukemia in 16 children. In these patients with late-stage disease, 10 responded with cancer remission and clinical improvements. The remissions, however, were temporary, and many fell ill while some had brief recoveries.[34] Nonetheless, it was a spectacular success—a true milestone that was unfortunately greeted with skepticism in the medical community when published in the *New England Journal of Medicine* in 1948—but widely acknowledged as a fundamental advance in cancer research.

At the same time that Farber was working on antifolates, including methotrexate, another strategy that achieved the same goal was being developed by George Hitchings and Gertrude Elion working on purine and pyrimidine analogs at Burroughs Wellcome laboratory in New York. Hitchings reasoned that these compounds could block DNA synthesis, which in turn might halt cell growth. Many scientists felt Hitchings' approach with these compounds was a shot in the dark, as the role of DNA in the workings of the cell was not well understood, and any link to cancerous growth was thought to be tenuous at best. Elion, an innovative synthetic chemist, had synthesized one purine antagonist, 2,6-diaminopurine, which showed promise in growth inhibition assays using bacterial cultures. In 1947, the team began sending Elion's compounds to the Sloan Kettering Institute, which had established drug screening for cancer chemotherapy agents in experimental models. Results in clinical trials with leukemic patients were promising, but the compound's toxicity was intolerable. Elion and Hitchings had developed two new chemotherapeutic drugs, thioguanine and 6-mercaptopurine by 1951. The 6-mercaptopurine drug was tried in patients who were resistant to methotrexate, and some 30 percent of those individuals responded with complete remission. The finding was soon confirmed, and 6-mercaptopurine as well as Farber's methotrexate were approved by the FDA in 1953 for treatment of acute lymphoblastic leukemia.

The trailblazing work on chemotherapeutics carried out by Elion and Hitchings introduced a more rational approach to drug discovery based on the understanding of basic physiochemical processes. More breakthroughs soon followed their foundational work. A biochemical insight from a rat tumor model led to yet another chemotherapeutic tactic tied to DNA and RNA synthesis. In the animal model, cancerous liver cells appeared to be exhausting the supply of uracil, one of the four bases of RNA, much more so than normal cells. A research group at the University of Wisconsin at Madison sought to exploit this difference by targeting the uracil biosynthetic pathway with uracil analogs. Charles Heidelberger synthesized 5-fluorouracil and then approached Hoffmann-La Roche, the pharmaceutical company in Nutley, New Jersey, to prepare a formulation for administering the drug to cancer patients with "far advanced malignant neoplastic disease."[35] The positive results propelled the drug to become the first chemotherapeutic agent against solid tumors and has remained an essential component of treatment regimens against colorectal cancer and many other cancers. Subsequent studies revealed that the compound's cytotoxicity is multifaceted. The drug is converted intracellularly into a substrate (5-fluorodeoxyuridine monophosphate, 5-FdUMP) for the enzyme thymidylate synthase and acts to inhibit production of deoxythymidine monophosphate (dTMP) needed for DNA synthesis and genome replication. In addition, 5-FdUMP is also surreptitiously incorporated into newly synthesized DNA and RNA, halting further macromolecular synthesis. Methotrexate's mechanism of action and pharmacology proved more difficult to unravel; it took 10 years to identify the drug's target, the enzyme dihydrofolate reductase, and another 50 years until a second class of antifolate drug, Eli Lilly's pemetrexed, was approved in 2004 for the treatment of mesothelioma.[36] Similar to 5-fluorouracil, methotrexate's utility lies in its ability to interfere with the biosynthetic pathways leading to DNA and RNA synthesis, depriving cells of essential building blocks and ultimately triggering cell suicide.[e]

Antibiotics as Cancer Chemotherapeutic Drugs

The chemotherapeutic race also left open a third lane to the scientific investigation of microbial compounds as a potential source of new drugs. Might bacteria produce compounds that are cytotoxic to proliferating human cells? Selman Waksman's research team at Rutgers University had isolated several antibiotic compounds from Actinobacteria and found one in particular, named Actinomycin A, that researchers at the Sloan Kettering Institute demonstrated was effective at inhibiting tumor growth in mice. In Germany, researchers at the Bayer Institute for Experimental Pathology (where Gerhard Domagk worked on antibiotics) had picked up on the news that such antibiotic compounds were capable of inhibiting tumor growth and found others, including Actinomycin C. Waksman tested the new version and also isolated Actinomycin D in 1953. Collaborating with Farber, the Actinomycin D version was shown to cure Wilm's tumor, a childhood kidney cancer, and it was later found to be effective in several other types of cancer. Two other antibiotics with similar cytotoxic activity were developed independently by laboratories in France, Italy, and Japan in the 1960s; these compounds were daunomycin and bleomycin.

SMALL MOLECULE DRUGS OF CANCER CHEMOTHERAPY

These are the small molecule drugs of cancer chemotherapy and their cellular actions:

DNA Damaging Compounds   DNA alkylating agents, including nitrogen mustards, cyclophosphamide, mitomycin, and cisplatin act primarily by chemically cross-linking of DNA strands. The cross-linking leads to attempts at DNA repair, cell cycle arrest, and the induction of a DNA damage response, triggering cell death via apoptosis.

Continues

---

[e] Methotrexate was discovered decades later to be an effective therapeutic for rheumatoid arthritis at vastly lower doses than used for chemotherapy; it was almost certainly the stigma of the drug's toxicity during high dose cancer treatments that kept it from being introduced as an anti-inflammatory medication.

DNA Synthesis Inhibitors   Compounds that inhibit DNA precursor molecule synthesis. Examples are the purine analogs methotrexate and 6-mercaptopurine, and the pyrimidine base analog 5-fluorouracil.

DNA Replication Inhibitors   Interference with the cell division cycle (anti-mitotic action) can be achieved with Vinca alkaloids such as vinblastine, Taxol, and DNA topoisomerase inhibitors such as camptothecin and daunorubicin.

RNA Transcription Inhibitors   The antibiotic drug actinomycin D interferes with all classes of RNA polymerases by intercalating with DNA and blocking polymerase progression.

The first generation of cancer therapeutics was built from discoveries and innovative treatment approaches made by academic scientists and clinical researchers working in collaboration with institute- or government-sponsored drug screening programs. The Sloan Kettering Institute became a focal point and offered services to academic labs as well as chemical and pharmaceutical companies. The National Cancer Institute (NCI) revived its drug screening program in the United States in 1955 as the Cancer Chemotherapy National Service Center (CCNSC) to build on the progress made in leukemias. In Europe, a similar initiative was started in 1957, known as the European Organization for Research and Treatment of Cancer. These drug screening initiatives played an important early role in encouraging pharmaceutical companies to take on the risks of drug development in treating cancer, in spite of market concerns and the often-dismal clinical response rates.

One of the darkest features of cancer is drug resistance, which develops almost invariably in all types of human malignant tumors. The genetic diversity found in tumors means that a subpopulation of cells (a subclone) will harbor a drug resistance mechanism, such as a way to pump the chemotherapeutic compound out of the cell via a transporter, allowing the subclone to grow unchecked. There is an array of mechanisms that present themselves under selective pressure of these toxic drugs. Not until the clinical development of effective combination therapies had been tested over decades of trials would cancers be effectively managed and pharmaceutical companies begin to see success.

Many other practical challenges remained in bringing a newly discovered drug candidate from a concept based on cell culture and animal model tests into the clinic. Mouse models in oncology were poor predictors of clinical outcomes, and how cancer cells usurped cellular growth controls was an enduring mystery well into the 1970s and 1980s.

Monoclonal antibody technology and the discovery of oncogenes and tumor suppressors by molecular biologists led the way to precision targeting of mutated signaling proteins driving uncontrolled cell proliferation. The development of Gleevec for targeting of a hyperactive kinase (due to a genomic rearrangement that produces the oncogenic Bcr-Abl fusion protein) in chronic myelogenous leukemia ushered in a new generation of drugs based on precision medicine.

Immunotherapy

For nearly a century, the paradigm for treating cancer was to focus on the use of chemotherapies and drugs that worked selectively on the cancer cells, feeding them cytotoxic compounds and targeting their dysregulated signaling pathways. Immunologists had puzzled over why cancers were so successful at evading the immune system, emerging as rogue residents yet able to thrive in the neighborhood. Researchers had long unraveled how T cells recognize foreign peptides on bacteria, viruses, or other sources of nonself biological material. The events underlying T cell signaling and activation were much more complex, as the stepwise process required engagement of multiple components to unleash an immune response. The discovery by James Allison and colleagues that a final step involved removal of a cellular checkpoint, or brake, was to completely upend the long-standing paradigm. Rather than treat the cancer, the researchers saw a way to treat the immune system instead.

The new approach was called *immunotherapy*. A T cell receptor called CTLA-4 acted as a brake, and antibodies that blocked its function could in turn cause T cell activation. Allison's team had presented the first evidence of anti-CTLA-4 antibodies working to block tumor growth in mice in 1996.[37] In humans, the first application of the checkpoint inhibitor ipilimumab (Yervoy, from Bristol-Myers Squibb) in metastatic melanoma led to long-term remissions (many years at least) in hundreds of patients whom otherwise would face an expected survival of 11 months. Later, a second checkpoint inhibitor had been identified by Tasuku Honjo and colleagues in Japan, a protein called PD-1 that is a receptor found on the T cell surface. Tumor cells were found to have the ligand, and its partner was termed PD-L1.[38,39] Following the FDA approval of Yervoy in 2011, Merck and Bristol-Myers Squibb developed monoclonal antibody drugs to independently target the PD-1 protein. Clinical results came quickly, and Merck's pembrolizumab (Keytruda) was approved in 2014. Keytruda ranks as the second highest in prescription drug sales in the United States at $11.1 billion in 2019. Bristol-Myer Squibb continued its immune-oncology success with nivolumab (Opdivo), a drug that ranked number four in sales in 2019, taking in $7.2 billion. The promise of checkpoint inhibitor drugs is their theoretically universal applicability towards any cancer type. This principle has led to an avalanche of activity in pharmaceutical and biotechnology companies to develop new immune-oncology drugs, particularly their use in combination. Incredibly, more than half of the 5,000 clinical trials underway in 2019–2020 were designed to test combination therapies across the vast landscape of cancer types and stages.

Figure 3.4 presents an overview of the history of cancer therapeutic development. The diagram links discoveries and innovations to the funnel of therapeutic modalities, highlighting the time lag—sometimes decades—between formation of an initial concept to arrival at an approved product. What is striking is the need for continuous innovation to bring new therapies online and how fundamental discoveries about the immune system have opened up an entirely new way to approach cancer in all of its deceptive and malignant forms.

The Pharmaceutical Business Model in the Twenty-First Century

Therapeutic discovery in present-day pharmaceutical companies traces its roots to the coal tar chemistry plants and local pharmacies established 150 years ago across Europe, the United Kingdom, and the United States. The principles of organic chemistry established and enabled the foundation for a new pharmacopeia based on the creation of synthetic or semi-synthetic compounds. What followed were extraordinary business opportunities for the coming century. In the pharmaceutical divisions of the rapidly diversifying chemical companies, the medicinal chemists and pharmacologists envisioned what could be accomplished with the new synthetic technology. Merck and Bayer were among the first to establish central research laboratories with a focus on process optimization. These companies relied on external innovation for initial products obtained with commercial arrangements from creative scientists such as Paul Ehrlich (Salvarsan, the anti-syphilitic drug), Martin Freund (Strypticin, a hemostasis drug), and Emil Fischer and Joseph von Mering, whom together made the first barbiturate, Veronal, a drug initially co-marketed by both Bayer and Merck. The opening of the antibiotic era with the discovery of microbial compounds also featured strong academic-industrial

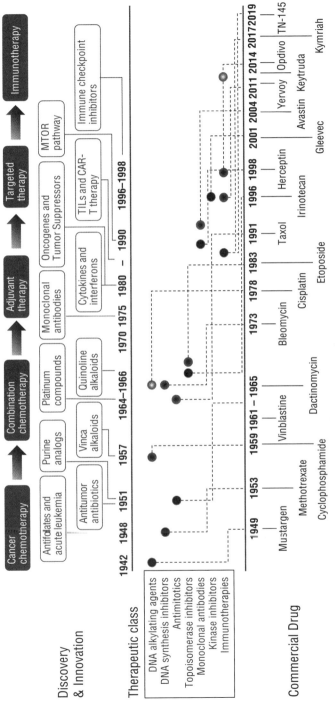

Figure 3.4: Cancer therapeutic development: linking innovation to industry

Discoveries and innovations bringing about the major advances in cancer treatment breakthroughs (shown on top portion of diagram) are shown in the upper frame with timeline; the first introduction of a chemical class or treatment modality is shown in the middle frame with a black circle, with second and third generations denoted with differently shaded circles. The link between therapeutic class or drug modality and eventual FDA product approval date is shown in the bottom frame, with the chemical or brand name of the product.

collaborations with Domagk, Gilman, Waksman, Farber, and many others. Post-World War II, most major pharmaceutical companies established microbiology departments and production facilities, and they acquired talent in bacteriology, pharmacology, and biochemistry to deepen their R&D capabilities. Today, these same pharmaceutical companies employ an army of chemists, translational medicine researchers, and disease biologists to unravel complex biology, which adds to the extraordinary expertise in therapeutic areas that has already been built up over long periods of time. Novartis, for example, has accumulated more than 50 years of experience in cardiovascular and psychiatric diseases, 50 years in immunology and 25 years' worth in oncology. This leads to the astonishing competitive advantages that these organizations now leverage to ensure a depth of new medicines and a scaled capability in their chosen therapeutic areas.

A critical piece of the modern drug development paradigm arose from industry's close relationships with hospital physicians and academic medical centers positioned at the core of healthcare systems. As illustrated by cancer drug development, physicians provided crucial contacts with patients and regulators and were key influencers of medical opinion. Institutes and clinicians at the forefront of medical oncology played a critical role in innovating around drug usage and new therapeutic modalities. Perhaps of greatest importance, advances made in clinical development were spurred on by the medical establishment and government regulators, which created the current framework of sponsor-investigator organized clinical trials.

The entire pharmaceutical enterprise hangs on the results from *clinical trials*; these are experiments in human populations with defined endpoints that provide executives, investors, and regulators with all-important, decision-making data. Increasingly complex clinical trial designs, including umbrella or basket trials, utilize precision medicine approaches that were brought forward in collaboration with clinicians. Precision medicine relies on molecularly defined *cohorts*—patient subpopulations defined by *biomarkers*. The FDA and other regulatory bodies are requiring demonstration of clinical efficacy of a candidate drug in a biomarker positive population and lack of efficacy in a negative population. These trials are invariably challenging when studies involve small groups and therefore have led drug companies to lean heavily on statistics, definitive genetic characterization by molecular testing, and the inclusion of real-world evidence generated from a variety of real-world data sources, such as sensor data from wearables, clinical laboratory data, and patient-reported outcomes.

The pharmaceutical industry's operational model has crystallized into a discrete set of activities from early discovery research to FDA approval, as outlined in Table 3.2.

Table 3.2: Discovery and Development of Small Molecule Drugs

STAGE	ACTIVITY	TASKS
Early Discovery 1–3 years	Target identification	▪ Assessment of existing or new biology on a disease target ▪ Omics analysis
	Target validation	▪ Target validation assay ▪ Genetic confirmation of disease association
	Hit discovery	▪ Compound library acquisition ▪ High throughput screening ▪ Structural biology ▪ Medicinal chemistry
	Hit confirmation	▪ Secondary assays for compound specificity, selectivity and mechanism ▪ Chemical class druggability
Preclinical Development 1–2 years	Lead identification	▪ Structure activity relationship (SAR) defined ▪ Synthetic feasibility ▪ *in vivo;* efficacy study ▪ Pharmacokinetics (PK) and toxicity studies
	Lead optimization	▪ Bioavailability, clearance, distribution, adsorption ▪ Formulation ▪ *In vivo;* PK and pharmacodynamics (PD) in-depth studies ▪ Biomarker discovery
	Development candidate	▪ Toxicity and drug interaction profiles ▪ GMP manufacture feasibility ▪ Clinical endpoint development ▪ Biomarker development
	Pre-IND and IND submission	▪ Acceptable clinical dosage and preclinical drug safety profile ▪ GMP quality ▪ IND regulatory path

Continues

Table 3.2 (*continued*)

STAGE	ACTIVITY	TASKS
Clinical Development 4–6 years	Human proof of concept (POC) Phase 1	■ Acceptable maximum tolerated dose and dose response ■ Evidence of human pharmacology ■ Safety assessment ■ Diagnostic development
	Clinical POC Phase 2	■ Acceptable PK/PD profile ■ Demonstration of efficacy ■ Biomarker data and patient stratification ■ Target engagement evaluation ■ Safety and tolerance
	Clinical POC Phase 3	■ Large clinical study ■ Efficacy relative to standard of care ■ Drug manufacturing ■ Companion diagnostic
Registration 0.5–1 year	Regulatory submissions	■ NDA or BLA filings ■ FDA reviews

Early discovery focuses on evaluating a therapeutic hypothesis based on biological knowledge or new research findings obtained by company scientists and collaborators. Targets that are identified in the process are validated and optimally only move forward with genetic confirmation of disease association. With small molecule discovery projects, a search commences for a chemical entity that can bind or associate with the target and affect biological activity. The hit discovery and confirmation activities utilize a variety of assays and high throughput screening technologies, ultimately leading to a decision point around nominating an early lead compound for preclinical development. The preclinical work focuses on absorption, distribution, metabolism, excretion, and toxicity (ADMET) studies, along with pharmacodyamic evaluation in animal models. In parallel, once feasibility of manufacture of a drug product with an active pharmaceutical ingredient is obtained, a decision is made whether to file an investigational new drug (IND) application. Post-IND approval, clinical development focuses on efficacy and safety in increasingly more complex trials, referred to as Phase 1, 2, or 3 clinical trials. (Phase 4, not included in Table 3.2, includes trials that are run on approved drugs. These post-marketing studies are for continued monitoring of drug safety and response.) Demonstration of drug efficacy and an acceptable safety profile requires a randomized, controlled, clinical trial (known as an RCT). The nuances of clinical trial design—its size, patient population, and duration reflect the drug product and target condition, as well as a roadmap to approval. The first RCTs were done in the 1940s and have become the gold standard for the unbiased, evidence-based information they produce. There are three important elements of any RCT that make it more likely to be definitive. These are comparing the experimental drug to a control; randomizing patients between the control and treatment groups; and, when possible and ethical, blinding the patients and clinicians as to whether patients are receiving the product being studied or the control.

R&D Productivity Challenges Within the Pharmaceutical Industry

The challenges and risks posed in discovering and developing new therapeutics as a business enterprise are unequaled and have few parallels in other industrial sectors. Several studies have explored product approval rates over time and find that on average about 10 percent of all INDs in the small molecule category eventually receive marketing approval from the FDA or the European Medicines Agency (EMA).[40-43] These are compounds that have already made their way as far as clinical evaluation in humans (Phase 1 or later, clinical development stage in Table 3.2).

There are manifold reasons for such a high attrition rate and the associated issues of high drug costs and stagnant R&D productivity across the industry. Drug safety and toxicity concerns account for a consistent, minor fraction of the attrition, anywhere from 10–30 percent across pharmaceutical companies. Even for projects where the biology is particularly well understood, the introduction of a novel small molecule drug or other drug types into humans may be harmful, toxic, or cause unacceptable side effects. In drug discovery and development operations, 70 percent of safety-related attrition occurs preclinically, meaning that 30 percent of drugs destined to fail due to safety issues are still evaluated in humans. Approximately 15 percent of total drug development costs are found within preclinical safety activities (Table 3.2). Small molecule drugs, even those possessing exquisite target specificity, may come into contact and interact with 100,000 different molecules in the human body.

Modeling such stochastic interactions and off-target effects is something that until recently has been extremely difficult to accomplish. Safety failures often arise in Phase 3, when a drug is tested in larger patient populations involving 1,000 or more participants compared to smaller Phase 1 and 2 trials enrolling only dozens to hundreds of volunteers, respectively. With larger population pools, a serious adverse event pattern can be uncovered, resulting in a late, and thus very costly, reason to halt a drug program. Clinical trial design, planning, and execution errors also contribute to overall failure by testing drugs in nonresponding or improperly stratified patient populations. Failures of this sort are due to drugs administered at the wrong time or wrong dose or without an understanding of human pharmacokinetics of a novel compound.

The main mode of failure in pharmaceutical development is lack of drug efficacy, as measured in clinical studies across Phases 1, 2, and 3. Analysis of 108 phase 2 trials in 2010 suggested that 50 percent of drugs fail in Phase 2 for efficacy, approximately 30 percent for "strategic" reasons, and 20 percent for safety and pharmacokinetic concerns.[44,45] Dissections of late-stage failures tend to point the finger at biology, where improperly vetted biological hypotheses and problems in translational research allow ill-fated molecules to make their way past earlier screening criteria and into human trials. These occur as the animal models or cellular systems that provide a biological basis for the intervention do not "translate" to human disease-modifying activity. Related to both the target and disease model, the target validation step appears to be the weak link, failing at linking the mechanistic hypothesis around the target to disease modification.[43] Another reason for failure is a suspected or confirmed lack of target engagement by the development candidate. These breakdowns come from projects where target-ligand engagement is not well-characterized or not demonstrated in living cells or tissues, versus *in vitro* biochemical assays, early in the development process. Costly projects also get shut down due to marketplace factors, for example when the efficacy profile of a drug is not deemed commercially competitive for filing a new drug application (NDA) at the regulatory review stage.

Despite a century of accumulated expertise and massive R&D expenditures by pharmaceutical companies, the price tag to develop new therapies has increased dramatically, rising 145 percent between 2003 and 2013 to $2.6 billion per new drug.[46] The industry might be running out of rope in seeking new blockbuster drugs that can be prescribed to vast markets for therapeutic maintenance of chronic conditions, such as diabetes, depression, or high cholesterol. There have been some obvious transition costs for switching from a business model that has historically been oriented around small molecule drugs to costlier biologics targeting genetically defined subpopulations. Greater challenges loom with pharmaceutical and biotechnology companies pursuing curative therapies, with medications costing hundreds of thousands of dollars per patient with smaller markets in rare diseases. The industry has also placed bets in areas of unmet medical needs where risks are higher.[47] To bear the ongoing technical risk and feed the blockbuster drug business model, any pharmaceutical company would need to possess a drug development pipeline where ten projects are ongoing, nine of which will ultimately fail along the way while they anticipate approval of a *new molecular entity* (NME) as a result of a single successful program. In reality, there are typically 25 or more projects grouped across several therapeutic areas (oncology, neuroscience, infectious disease, as examples) that are moved forward in parallel.

Building a pipeline of the magnitude described and sustaining the operational and competitive advantages that come with scale require enormous capital for R&D investment, manufacturing facilities, and sales and marketing activities to achieve significant returns for the company and its shareholders. Big Pharma companies have colossal, multibillion-dollar R&D budgets to innovate, drive drug development pipelines, and maintain competitiveness in their chosen therapeutic areas. In 2017, the five largest pharmaceutical companies—Roche, Johnson & Johnson, Pfizer, Merck, and Novartis—each reported R&D expense of somewhere between $7.5 (Novartis) and $12.5 billion (Roche), and together the companies outspent the entire budget of the United States National Institutes of Health, which was $34 billion.

The number of new drugs entering the market has remained fairly constant over the past several decades, in spite of massive increases in biomedical knowledge, pharmaceutical manufacturing expertise and scale, computational resources, and innovative chemistry.[48] The entire drug industry annually produces on average 31 novel FDA approvals per year, as illustrated in Figure 3.5. From 1993 to 1999, the average was 33; a trough of low productivity was seen over the years 2000 to 2013, when the average dipped to 25 per year. Over 2014 to 2019, the trend was upwards, with an average of 43 new approvals. The question that continues to linger is: Why is there not a steady increase given the resources poured into drug development?

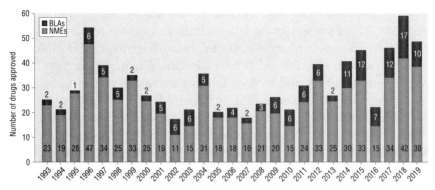

Figure 3.5: Novel FDA approvals since 1993

Annual numbers of new molecular entities (NMEs) and biologics license applications (BLAs) approved by the FDA's Center for Drug Evaluation and Research (CDER). The chart excludes vaccines and other biologics approved by the Center for Biologics Evaluation Research.
Source: Drugs@FDA. Figure reproduced with permission from Mullard, A. 2019 FDA drug approvals, Nature Reviews Drug Discovery. 19(2):79–84; 2020.

The most glaring feature of drug development from a business perspective is the technical risk that is encountered on the road to a new product. It is instructive to compare the pharmaceutical industry to other major industrial sectors, such as energy or technology. Oil and gas exploration may be fraught with risks, with investors gambling on oil field location and size, but the natural resource will be found, somewhere, with absolute certainty. The technical risk is very low, although supply and demand—the market risk—is higher. The only other sector with comparable metrics to pharmaceuticals is found in technology, where semiconductor companies allocate 20 percent of total revenues to ongoing R&D. Manufacturing of semiconductor integrated circuits has followed Moore's law, which at its core is a business model that has been followed for five decades. It is driven by engineering and innovative process improvements, yielding compute devices that are twice as powerful and delivered at lower cost relative to performance with each successive generation.

The most advanced process nodes, such as those at 7nm, require engineering of physical components at near atomic precision. The chip design cost for the 7nm node was estimated at $300 million, on par with development costs for an NME when not factoring in other drug program failures. Again, although there is technical risk, this is drastically lower when compared to drug development. R&D problems in semiconductor manufacturing wind up slowing down cycle time, as was seen in 2020 with Intel's delay of its 7nm chips to 2022. Up until recently, successful roadmaps for moving to smaller nodes were planned years in advance. TSMC, the giant Taiwanese manufacturer, brought its 7nm node production online in 2018 following a long line of improvements since 1997, when the 250nm process node was first introduced. The drug industry has nothing comparable to this built-in success planning, and it has not adopted an end-to-end engineering approach to solve its R&D productivity issues.

How can operational efficiency and R&D productivity be increased at the organizational level? The economics of drug development force managers to maximize both R&D productivity and a drug's value, while lowering the cost per new drug (improving operational efficiency). This can be formalized in an equation relating productivity to a set of the previous elements and other key factors affecting output, as follows:

$$P = (WIP \times pTS \times V) / (CT \times C)$$

In the equation, P is productivity, WIP is work in process, pTS is the probability of technical success, V is the portfolio value, CT is the cycle time, and C is the cost.[45] Management and R&D's goals are thus to find ways to increase the volume of innovation (the WIP, or number of NMEs in development) and the probability of technical success and value, and then attempt to lower project cycle time and cost. A look at AstraZeneca's productivity over multiple time spans indicated that changes in the quality of WIP added significantly to their productivity gains.[43] One of the areas where productivity gains are most evident is in the selection of therapeutic modality. The success rate of large molecule drug discovery, those known as *biologic* drugs, is more than double that (25 percent) found for small molecules (10 percent). The industry's long, painful experience with dismal success rates for small molecule cancer therapeutics (see Figure 3.5, the first three chemotherapeutic classes identified) has been followed by a headlong rush into immunotherapies—a harbinger of how the industry will adapt to tackle the productivity equation.

Sources of Pharmaceutical Innovation: Biotechnology and New Therapeutic Modalities

Biotechnology companies are the primary source of new product and technology innovation across the biopharmaceutical industry. The biotechnology revolution in the late 1970s and 1980s set in motion the third phase of therapeutic discovery, with early pioneers such as Genentech, Amgen, Biogen, and Immunex establishing top-tier research laboratories from the outset. Molecular biologists led the charge in using recombinant DNA technology to create biotherapeutic drugs and increased the number of addressable targets by an order of magnitude. At the end of the twentieth century, small molecule drugs addressed about 500 molecular targets. Of these, nearly half were membrane proteins known as G-protein coupled receptors, and the remainder were enzymes—ion channels and transporters.[49] The target landscape is now vastly richer with the tools of molecular biology and new drug modalities, enabling effective targeting of soluble growth factors, intracellular signaling molecules, cell-cycle proteins, transcription factors, apoptosis modulators, and many other types of targets. Figure 3.6 illustrates the expansion of therapeutic (drug) modalities beyond small molecules.

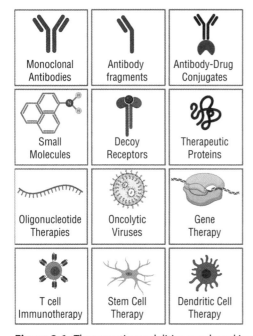

Figure 3.6: Therapeutic modalities employed in current drug development pipelines

Biotechnology has enabled the development of several novel therapeutic modalities, including full-length monoclonal antibodies, antibody fragments, and antibody-drug conjugates (top row); two classes built from recombinant DNA technology are decoy receptors and therapeutic proteins (second row); another group based on delivery of genetic information into cells are oligonucleotide therapies, oncolytic viruses, and gene therapy (third row); and cell-based therapies include those that can be genetically engineered as in T cell immunotherapy (also known as CAR-T), or as re-programmed or purified cells, including stem cell and dendritic cell therapy (fourth row).

Antibody-based therapies in particular have dramatically changed the landscape of drug development. Approximately 100 monoclonal antibody drugs have been approved since the pioneering muromonab-CD3 was approved by the FDA in 1986. Half of these drug approvals occurred after 2008, primarily for oncology and immune system disorders. Today, seven of the top ten best-selling drugs in the United States are monoclonal antibodies. The incredible target specificity and wide applicability of monoclonal antibodies has established this therapeutic modality as a favored starting point for drug development.

Large pharmaceutical companies historically have lagged well behind biotechnology firms in terms of innovation and have boosted their own R&D productivity principally through partnerships and acquisitions. The merger and acquisition more than of Big Pharma are well documented. From 1998 to 2012, over $300 billion was invested in acquisitions of 132 biotechnology companies.[50] This financial frenzy resulted in a massive restructuring of the biotech industry, leaving only Amgen, Gilead, and Regeneron remaining of the top 10 companies. The pattern of financial investments indicates that the larger pharmaceutical companies wait for clear indications of scientific value and business opportunity prior to moving into a new therapeutic arena. Immuno-oncology is a perfect example, where investment prior to Yervoy's approval in 2011 was rather restrictive. Bristol-Myers Squibb made a bet as a first mover, and it paid off. The company acquired Medarex Inc in 2009 for $2 billion, from which came both Yervoy and Opdivo. Medarex had developed a "fully human" antibody platform in transgenic mice that were engineered with components of a human immune system to produce antibodies as drugs.

Innovation is the lifeblood of the biopharmaceutical industry. Big Pharma companies will leverage their financial resources, advantages of scale, and operational expertise to improve productivity and bring new medicines to market. Biotechnology companies will continue to innovate and obtain financial investments from public markets, venture capital, pharmaceutical partnerships, and corporate venture investments to build new tools and therapeutics. The base of innovation is poised to expand, as new technologies from adjacent fields find valuable applications in biology and medicine. Chief among these will be molecular scale engineering and artificial intelligence. New medicines are beginning to emerge from the fourth wave of therapeutic discovery, as research and clinical teams design and engineer purpose-built therapeutics for cancer, rare disease, and longevity.

EXPANDING THE INNOVATION BASE IN PHARMACEUTICAL ENTERPRISES

The following are ways to expand the innovation based in pharmaceutical enterprises:

Synthetic Chemistry Innovation   A powerful example is ring-closing metathesis chemistry that has enabled the discovery of six approved Hepatitis C drugs: simeprevir, paritaprevir, grazoprevir vaniprevir, voxila-previr, and glecaprevir. The novel chemistry allows for assembly of complex, bioactive molecules for difficult targets.

Natural Product Screening   Approximately 40 percent of new small molecule drugs in the past 25 years have originated from the discovery of secondary metabolites in plants and microbes. The structural complexity of molecules found in living organisms is far higher than that which can be typically achieved by chemical synthesis. Natural products harbor an average of 6.2 chiral centers per molecule, as compared to an average of 0.4 chiral centers found in combinatorial libraries. The chemical space remains mostly untapped, as only 15 percent of the planet's 350,000 species or millions of microbes have been characterized at the level of their biochemical pathways and chemical constituents.

Network Pharmacology and Polypharmacology   A new approach put forward that is aimed at multitarget drug mechanism of action. The concept is to go beyond developing drugs around single targets and seek a pattern of targets.

Computational Chemistry, Machine Learning, and Quantum Computing   Continued adoption of computational techniques has the potential for new reaction discovery and optimization, catalyst design, and reaction prediction. Quantum computing has arrived, and it will soon enable prediction of molecular properties and chemical reaction simulations.[51] New modes of molecular editing will emerge to selectively insert, delete, or exchange atoms, analogous to what is done with gene editing technologies for DNA and RNA molecules.

Genetic, Cell, and Metabolic Engineering   The future of medicine may lie in the industry's technical ability to engineer cells and genes for curing cancer and rare genetic diseases. Metabolic engineering in microbes to produce human therapeutics introduced into the gut microbiome is already underway.

Artificial Intelligence in Discovery Research and Early Translational Medicine   AI-based decision-making, risk analysis, and hypothesis testing during the early stages of discovery research could bring together an understanding of drug properties and disease intervention needs amidst uncertainty.

Notes

1. Prüll C-R, Maehle A-H, Halliwell RF. Drugs and Cells—Pioneering the Concept of Receptors. Pharm Hist. 2003;45(1):18–30.

2. Langley JN. On the reaction of cells and of nerve-endings to certain poisons, chiefly as regards the reaction of striated muscle to nicotine and to curari. J Physiol. 1905;33(4–5):374–413.

3. Frenzel A, Schirrmann T, Hust M. Phage display-derived human antibodies in clinical development and therapy. MAbs. 2016 Jul 14;8(7):1177–94.

4. Salavert A. Agricultural Dispersals in Mediterranean and Temperate Europe. In: Oxford Research Encyclopedia of Environmental Science. 2017.

5. Bakels, C. C. Fruits and seeds from the Linearbandkeramik settlement at Meindling, Germany, with special reference to Papaver somniferum. ANALECTA Praehist Leiden. 1992;25:55–68.

6. Inglis, Lucy. Milk of Paradise: A History of Opium. New York London: Pegasus Books; 2018.

7. Martin L. Plant economy and territory exploitation in the Alps during the Neolithic (5000–4200 cal bc): first results of archaeobotanical studies in the Valais (Switzerland). Veg Hist Archaeobotany. 2015 Jan;24(1):63–73.

8. McIntosh, Jane. Handbook of Life in Prehistoric Europe. Oxford University Press USA; 2009.

9. Brenan, Gerald. South from Granada. Penguin Press; 2008.

10. Harding, J., Healy, F. A Neolithic and Bronze Age Landscape in Northamptonshire: The Raunds Area Project. English Heritage; 2008.

11. Kapoor L. Opium Poppy: Botany, Chemistry, and Pharmacology. CRC Press; 1997.

12. Scholtyseck J, Burhop C, Kibener M, Schafer H. Merck: From a Pharmacy to a Global Corporation. 1st ed. C.H. Beck; 2018.

13. Smith A. An Inquiry into the Nature and Causes of the Wealth of Nations. Vol. I. London: W. Strahan and T. Cadell; 1776.

14. Holloway SWF. The Apothecaries' Act, 1815: A Reinterpretation. Med Hist. 1966;10(2):107–29.

15. Sneader W. The discovery of aspirin: a reappraisal. BMJ. 2000 Dec 23;321(7276):1591–4.

16. Hirsch D, Ogas O. The Drug Hunters: The Improbable Quest to Discover New Medicines. Arcade; 2016.

17. Plater MJ. WH Perkin, Patent AD 1856 No 1984: A Review on Authentic Mauveine and Related Compounds. J Chem Res. 2015 May 1;39(5):251–9.

18. Drews J. Drug Discovery: A Historical Perspective. Science. 2000 Mar 17;287(5460):1960–4.

19. Lerner J. The Architecture of Innovation: The Economics of Creative Organizations. Harvard Business Review Press; 2012.

20. Kekule A. Sur la constitution des substances aromatiques. Bull Soc Chim Fr. 1865;3(2):98–110.

21. Amyes SJB. Magic Bullets: Lost Horizons: The Rise and Fall of Antibiotics. London: Taylor & Francis, Inc.; 2001.

22. Bosch F, Rosich L. The Contributions of Paul Ehrlich to Pharmacology: A Tribute on the Occasion of the Centenary of His Nobel Prize. Pharmacology. 2008 Oct;82(3):171–9.

23. Yeadon G, Hawkins J. The Nazi Hydra in America: Suppressed History of a Century. Progressive Press; 2008.

24. Doctors Turn Over $500,000 to Science. Profits From Wartime Sales of Their Substitute for Salvarsan. New York Times. 1921 Mar 3;

25. Church R, Tansey T. Burroughs Wellcome & Co.: knowledge, trust, profit and the transformation of the British pharmaceutical industry, 1880-1940. Lancaster, UK: Crucible Books; 2007.

26. Renwick G. German Dye Trust to Fight for World Trade. New York Times. 1919 Dec 3;

27. Taussig FW. Germany's Reparation Payments. The Atlantic. 1920 Mar;

28. Pope WJ. Synthetic Therapeutic Agents. Br Med J. 1924 Mar 8;1(3297):413–4.

29. Fleming A. On the antibacterial action of cultures of a Penicillium with special reference to their use in the isolation of B. influenza. Br J Exp Pathol. 1929;(10):226–36.

30. Chain E, Florey HW, Adelaide MB, Gardner AD, Oxford, DM, Heatley NG, et al. Penicillin as a chemotherapeutic agent. The Lancet. 1940;236(6104):226–8.

31. Abraham EP, Chain E, Fletcher CM, Florey CM, Gardner AD, Heatley NG, et al. Further observations on penicillin. The Lancet. 1941;238(6155):177–89.

32. Krumbhaar EB, Krumbhaar HD. The Blood and Bone Marrow in Yellow Cross Gas (Mustard Gas) Poisoning: Changes Produced in the Bone Marrow of Fatal Cases. J Med Res. 1919;40(3):497–508.

33. Bordin DL, Lima M, Lenz G, Saffi J, Meira LB, Mésange P, et al. DNA alkylation damage and autophagy induction. Mutat Res. 2013 Dec;753(2):91–9.

34. Farber S, Diamond LK, Mercer RD, Sylvester RF, Wolff JA. Temporary remissions in acute leukemia in children produced by folic antagonist, 4-aminopteroylglutamic acid (aminopterin). N Engl J Med. 1948;238:787–93.

35. Curreri AR, Ansfield FJ, McIver FA, Waisman HA, Heidelberger C. Clinical Studies with 5-Fluorouracil. Cancer Res. 1958 May 1;18(4):478–84.

36. Jolivet J, Cowan KH, Curt GA, Clendeninn NJ, Chabner BA. The pharmacology and clinical use of methotrexate. N Engl J Med. 1983 Nov 3;309(18):1094–104.

37. Leach DR, Krummel MF, Allison JP. Enhancement of Antitumor Immunity by CTLA-4 Blockade. Science. 1996 Mar 22;271(5256): 1734–6.

38. Nishimura H, Minato N, Nakano T, Honjo T. Immunological studies on PD-1 deficient mice: implication of PD-1 as a negative regulator for B cell responses. Int Immunol. 1998 Oct 1;10(10):1563–72.

39. Nishimura H, Nose M, Hiai H, Minato N, Honjo T. Development of Lupus-like Autoimmune Diseases by Disruption of the PD-1 Gene Encoding an ITIM Motif-Carrying Immunoreceptor. Immunity. 1999 Aug 1;11(2):141–51.

40. Hay M, Thomas DW, Craighead JL, Economides C, Rosenthal J. Clinical development success rates for investigational drugs. Nat Biotechnol. 2014 Jan;32(1):40–51.

41. KMR Group. Pharmaceutical Benchmarking Forum. Probability of success by molecule size: large vs small molecules. Chicago: KMR Group; 2012.

42. DiMasi JA, Chakravarthy R. Competitive Development in Pharmacologic Classes: Market Entry and the Timing of Development. Clin Pharmacol Ther. 2016;100(6):754–60.

43. Morgan P, Brown DG, Lennard S, Anderton MJ, Barrett JC, Eriksson U, et al. Impact of a five-dimensional framework on R&D productivity at AstraZeneca. Nat Rev Drug Discov. 2018 Mar;17(3):167–81.

44. Arrowsmith J, Miller P. Phase II and Phase III attrition rates 2011–2012. Nat Rev Drug Discov. 2013 Aug 1;12(8):569–569.

45. Paul SM, Mytelka DS, Dunwiddie CT, Persinger CC, Munos BH, Lindborg SR, et al. How to improve R&D productivity: the pharmaceutical industry's grand challenge. Nat Rev Drug Discov. 2010 Mar;9(3):203–14.

46. DiMasi JA, Grabowski HG, Hansen RW. Costs of developing a new drug. Tufts Center for the Study of Drug Development Briefing. Innovation in the pharmaceutical industry: new estimates of R&D costs. 2014.

47. Pammolli F, Magazzini L, Riccaboni M. The productivity crisis in pharmaceutical R&D. Nat Rev Drug Discov. 2011 Jun;10(6):428–38.

48. Newman DJ, Cragg GM. Natural Products as Sources of New Drugs over the Nearly Four Decades from 01/1981 to 09/2019. J Nat Prod. 2020 27;83(3):770–803.

49. Drews J, Ryser S. The role of innovation in drug development. Nat Biotechnol. 1997 Dec;15(13):1318–9.

50. Evens RP, Kaitin KI. The biotechnology innovation machine: a source of intelligent biopharmaceuticals for the pharma industry-mapping biotechnology's success. Clin Pharmacol Ther. 2014;95(5):528–32.

51. Google AI Quantum and, Arute F, Arya K, Babbush R, Bacon D, Bardin JC, et al. Hartree-Fock on a superconducting qubit quantum computer. Science. 2020 Aug 28;369(6507):1084–9.

Recommended Reading

Drews, Jürgen. In Quest of Tomorrow's Medicines. New York: Springer-Verlag; 1999.

Hirsch, Donald and Ogas, Ogi. The Drug Hunters: The Improbable Quest to Discover New Medicines. New York: Arcade; 2016.

Mukherjee, Siddhartha. The Emperor of All Maladies: A Biography of Cancer. Scribner; 2010.

Graeber, Charles. The Breakthrough: Immunotherapy and the Race to Cure Cancer. New York: Twelve, Hatchett Book Group, Inc.; 2018.

Gene Editing and the New Tools of Biotechnology

This system offers a straightforward way to cleave any desired site in a genome, which could be used to introduce new genetic information by coupling it to well-known cellular DNA recombination mechanisms.

Jennifer Doudna, after publication of the CRISPR breakthrough in *Science*, 2012

We have seen so many scientists and developers immediately embrace the technology and apply it to their own needs to perform gene editing and engineering of cells and organisms. The technology has already had a huge impact in life sciences, biotechnology, and biomedicine with the technology enabled to treat patients with serious blood diseases.

Emmanuelle Charpentier, personal communication, December 2, 2020

A generation after DNA's double helical structure was determined in 1953, biotechnology arrived via academic research laboratories that were developing tools to manipulate genes in living organisms. Molecular biologists had discovered an array of enzymes in bacteria that could cut DNA at sequence-specific locations, glue double-stranded pieces together, and synthesize copies in a test tube. Although these scientists were addressing basic research problems, they had also created the essential laboratory methods that comprised the foundation for the first generation of tools used in biotechnology. From the small cadre of molecular biologists in the 1960s and 1970s who learned the black art of gene cloning, a whole generation of scientists flourished, nearly all with pedigrees tracing back to the hippie scientists and seminal figures that founded the discipline. In many ways, it was an echo of other tech revolutions. Once the lab freezers were stocked with precious enzymes and cloning vectors, as the protocols for gene splicing in bacteria were written down, tested, and validated, they all became as sharable as software. The culture in molecular biology labs was not unlike the one shared between the hackers and software developers scattered across dorm rooms and offices in corporate America, which sprung up in parallel from the late 1970s and still persists today.

Another remarkable feature, and an enduring hallmark of the development of recombinant DNA technologies, is that they are among the most inexpensive, versatile, and low-tech tools in the world of science. The fundamental discoveries uncovered with the application of biotechnology did not require billion-dollar telescopes, supercomputers, or particle accelerators to peer into the workings of the cell—nor were teams of theorists needed to unravel the data. All that was required was careful genetic and biochemical experimentation, a few well-chosen microbes, large numbers of petri dishes and test tubes, and a creativity and curiosity for understanding processes essential for life. Creating new versions of genes in bacteria and then shuttling them into mammalian cells allowed scientists to derive molecular rules governing the cell cycle, oncogenesis, DNA replication, RNA transcription, and protein synthesis—along with linking observations of newly discovered genomic elements with predictions of their biological function.

The formation of biotechnology companies demarcated a new era of therapeutic discovery based on genetic engineering. The very first patent application for recombinant DNA technology was filed by Stanford University in 1974. After a long journey through the US Patent and Trademark Office, the patent was eventually granted in 1980—a harbinger of the fights to come over commercial ownership of human genes, which was ultimately decided against by the Supreme Court on June 13, 2013. The initial promise of the biotech revolution was to deliver native biological molecules like human insulin as drugs. For the pharmaceutical industry, this was foreign ground, and they had envisioned a long maturation period. But the youthful biotechnology startups proved them wrong, and success arrived more quickly than perhaps any other drug discovery technology in history. From the founding of Genentech in 1976 until the approval of its first drug (licensed to Lilly), a mere six years had elapsed. It became abundantly clear that medicine would not have to wait a generation to see results from biotechnological innovation. Soon, gene hunting pioneers and medical geneticists expanded the vision and embarked on a path toward gene therapy—seeking to use recombinant DNA technology to introduce new genetic material into human cells to overcome deficiencies or to replace defective genes in serious genetic conditions.

The first clinical trials of gene therapy were initiated in 1990, and hundreds more were approved to move forward by the FDA up until the death of a clinical trial participant, Jesse Gilsinger, in 1999.[1] There was enormous hype coupled with serious concerns about safety of the viral delivery systems being used in gene therapy and the dangers of random integration of extrachromosomal DNA into a human genome. While eagerly anticipated to follow closely on the heels of the first biotech drugs, the gene therapy field stood still, grappling with the difficulties of gene delivery and the continued occurrence of serious adverse events, which included incidences of retroviral vector-linked cancer, in clinical trials.[2,3] Following a decade's worth of intensive research to design safer vectors, the first gene therapy approvals in the Western world came about only in 2017—consuming a generation's worth of therapeutic ambitions to target the genome.[4]

At the turn of the twenty-first century, recombinant DNA tools for genome engineering were rather crude by today's standards. Restriction enzymes that were the bread and butter of molecular biologists did not have sufficient specificity to target the human genome at sites bearing disease-causing mutations. Gene therapy was done using techniques to transfer entire genes into cells *ex vivo* or to use viruses for *in vivo* delivery of genetic material encoding replacement genes. No technologies existed to edit the genome precisely without unintended large-scale genomic changes or off-target editing elsewhere. How were scientists and clinicians going to guide a molecular machine with laser-like accuracy to a location on a genome and also perform precision chemistry on single nucleotide variants? The solution would arrive from processes developed over a billion years of coevolution and by modern protein engineering in biotechnology labs.

Not unsurprisingly, the source of the second-generation biotechnology tools had their origins in prokaryotes. Researchers examining microbial genomes uncovered a class of bacterial defense systems designed with properties of a mammalian adaptive immune system, but with a memory and recognition capability for DNA and RNA. The unexpected discovery of the CRISPR-Cas system and the insight by Doudna and Charpentier in 2012 that the molecular complex could be reconfigured and repurposed as a programmable genome editor was the launch of a world-changing second generation of biotechnology tools.[5,6]

Unlike the sparse beginnings of biotechnology, the second wave was able to leverage all of the computational advances, accumulated data, sequencing technology, and molecular cloning toolkit of the previous generation to enable biology to enter an era of genome engineering. The rapid acceleration into this era began at the inflection point where CRISPR-Cas systems were envisioned, as illustrated in Figure 4.1.

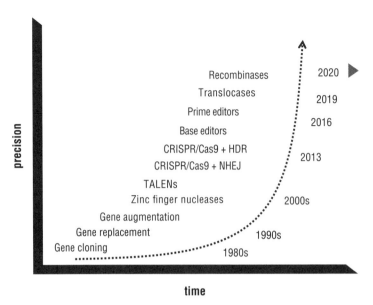

Figure 4.1: Biotechnology tools and the acceleration into an era of precision genome engineering

The tools of recombinant DNA technology opened the door to the possibility of genetic manipulation of genes, traits, and organisms. The early tools and techniques (1980–1990s) could recombine and transfer gene segments into cells, including embryos, but lacked precision. Second-generation biotechnology tools based on CRISPR-Cas led to a dramatic shift in precision due to the nature of programmable nucleases. Continued research and development to engineer more efficient and precise editing technologies has led to dramatic leaps in the breadth of genomic edits possible with relatively few components: a programmable RNA guide and a multifunctional protein built on a Cas scaffold that enables high precision genome editing for an array of applications. TALENs: transcription activator-like effector nucleases; ZFNs: zinc finger nucleases; NHEJ: non-homologous end joining; HDR: homology-directed repair.

Biologists began to exploit CRISPR-based gene targeting tools for what has become a long laundry list of applications. One of the first uses began with genetic modification of prokaryotes. There is now an enormous catalog of industrial microbes whose genomes have been engineered with CRISPR-Cas systems to produce drugs, biochemicals, and biofuels more efficiently.[7] To study gene regulation and function in eukaryotes, the CRISPR-Cas system was redesigned with the ability to turn genes on and off with transcriptional modifiers and also engineered with the ability to perform gene knockouts and knock-ins. The biotech tools of the CRISPR era are versatile, inexpensive, and even more democratized than their recombinant DNA ancestors, especially when it comes to transgenic animal applications. Researchers are now able to use CRISPR to engineer new plant and animal models in a few weeks' time to explore the functional consequences of single-base mutations in one or more genes, a process that had previously taken up to a year in some species with long reproductive cycles. Beyond basic research, CRISPR approaches are being used in agricultural biotechnology to engineer crops and livestock more easily for important agricultural traits and disease resistance. The molecular diagnostic capabilities of CRISPR were on display during the worldwide effort to control the SARS-CoV-2 pandemic, with assays from Sherlock Biosciences and also Mammoth Biosciences and UCSF (DETECTR) that provided rapid turnaround alternatives to detect the virus from RNA samples.

In medicine, precision genome engineering has long been considered a Holy Grail and medicine's future—a technology that would finally deliver permanent, curative therapies for many rare genetic diseases and some forms of cancer. With the lessons of gene therapy behind it, would CRISPR-based approaches be able to overcome the inherent risks and limitations in editing human genomes? What about germline editing for human embryos? Assisted reproductive technologies were in use well in advance of genome engineering capabilities. The simplicity of CRISPR-Cas systems that need only two components—the guide RNA and Cas protein—created the possibility that rogue actors might deploy embryo editing before the technology, medical science, and society were ready. Indeed, the field was horrified but not entirely blindsided by the first attempt at heritable human genome editing that was announced in 2018 by Chinese scientist Jiankui He. The eventual birth of twins with edits made to their CCR5 gene (in theory to disable the CCR5 cell surface receptor as a prophylactic against HIV infection) was denounced as unethical and medically inappropriate—and potentially harmful. Criticism was swift and universal with appeals for a moratorium on any heritable human gene editing. A glaring spotlight was also shone on the ethics, medical needs, and dangers of genome editing at this stage of CRISPR-Cas's technological evolution and maturation. Fortunately, for second-generation gene therapy, much more promising and relatively safer applications are making their way through clinical trials for tackling blood disorders, such as sickle cell disease and certain hematologic malignancies, as well as cystic fibrosis, muscular dystrophy, and inherited retinal dystrophies.

The CRISPR breakthrough in 2012 and ongoing innovation in the academic sector led to a whole new ecosystem of biotechnology companies. The first graduating class were startups focused either on therapeutics or on supplying tools for the burgeoning set of CRISPR applications. The frontrunners in therapeutics were those cofounded by CRISPR-Cas system pioneers: Nobel laureates Jennifer Doudna (Intellia Therapeutics and Caribou Biosciences) and Emmanuelle Charpentier (CRISPR Therapeutics, ERS Genomics) in addition to George Church from Harvard University and Feng Zhang at the Broad Institute (Editas Medicine).[*] Billions were poured into these and other companies from private venture capital, public markets, and the pharmaceutical industry in attempts to commercialize the CRISPR discoveries. The questions for the biotechnology sector were many of the same longstanding ones for gene therapy. For the technology, how does the delivery problem get solved? Can durable and permanent edits be achieved? Can gene editing be made safe and effective for a wider range of diseases? If these enduring problems can be solved, then genome engineering will enable a new frontier of medicine. Remarkably, this will be accomplished with simple tools derived from microbes and ancient bacterial defense systems relying on the universal code of life and the storage of biological information in DNA.

Molecular Biology and Biological Information Flow

Information storage in DNA is fundamental and shared by all life forms, providing nature's instruction manual for the development, replication, and survival of complex adaptive organisms. Deciphering an organism's genome only became possible after understanding the information flow that became known as the "central dogma" of molecular biology. Following the discovery of the double helical structure of DNA by James Watson and Francis Crick in 1953, the major question was: How is information encoded in DNA? Or more precisely, how do genes make proteins? More than a decade earlier, George Beadle and Edward Tatum had already proved the hypothesis that there was a one-to-one correspondence of gene to protein. If this were true, would a linear sequence of nucleotides in DNA then specify the linear sequence of amino acids in proteins? To add to the complexity, there were indications that RNA molecules were

---

[*]Jennifer Doudna was also a cofounder of Editas; however, she left the company after a patent dispute between UC Berkeley (her academic affiliation), and the Broad Institute was decided at the USPTO in favor of Zhang's earlier date of invention using human cells.

present in cells and might be involved in protein synthesis. What role did RNA play? The beginning conversations of how the information transfer could be accomplished were carried on by Watson, Crick, and others in a small circle of scientists in Cambridge, England. Francis Crick had famously laid out a conceptual model of the central dogma in a schematic cartoon, illustrating the idea even though he had not determined the type of RNAs involved in the process or indeed how it all worked, as shown in Figure 4.2. All of these questions were the basis of momentous experiments to finally unravel the genetic code.

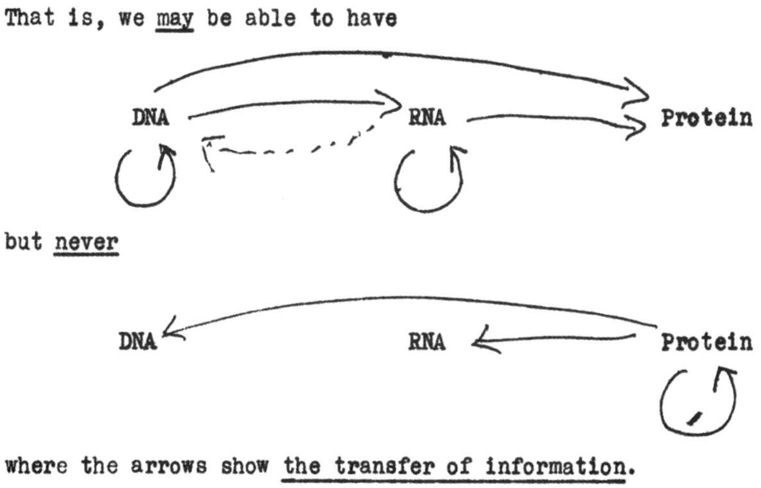

Figure 4.2: The first illustration of the "central dogma," as drawn by Francis Crick in 1956

Image: Wellcome Library, London

Research in the biological sciences in the last half of the twentieth century involved laborious, small-scale laboratory research without the help of computers. For the scientists involved in the new field of molecular biology (also called *biochemical genetics* or *chemical genetics* at the time), this meant purifying enzymes, synthesizing radioactive tracer chemicals, developing repeatable laboratory assays, and testing a handful of conditions in carefully constructed experiments over days, weeks, or even months. The results of experiments were kept in small notebooks and X-ray films, with data points numbering in the few hundreds at the extreme. The pace of progress suffered not from lack of conceptual understanding (or intellectual ability!) but from limited technologies and instrumentation with which to generate experimental data. Nonetheless, the enormous challenge spurred a wonderful era of scientific creativity.

Putting together the intricacies of cellular biochemistry required simpler systems in test tubes where cellular extracts and purified components could be combined and chemical reactions evaluated. The development of these *in vitro* methods was critical for piecing the puzzle together. Hypotheses had to be tested, and ideas either discarded or kept, with the winners tested further to make progress. Francis Crick had suggested that perhaps there were adapter molecules that linked amino acids to the information contained in DNA. Over time, in various laboratories in the United States and Europe, evidence had accumulated to suggest that he was on the right track. At Massachusetts General Hospital in Boston, Paul Zamecnik and Mahlon Hoaglund demonstrated that protein synthesis occurred on ribosomes in the cytoplasm, outside of the nucleus. Ribosomes are comprised of RNA subunits, but these were not suitable to convey genetic information. Soon thereafter, these researchers discovered another class of RNA molecule, known as transfer RNAs (tRNAs), that were bound (chemically linked) to amino acids.[8] If this RNA contained complementary nucleotide sequences, maybe they could pair with DNA or RNA sequences. If so, they could link DNA instructions to the protein synthesis machinery. So then, where was the template for directing protein synthesis? If the process was taking placing in a separate cellular compartment away from DNA in the nucleus, what conveyed the signal?

The answer came in a series of experiments demonstrating that messenger RNA was the intermediary between DNA and protein. The titles of the papers published simultaneously on May 13th, 1961, in the journal *Nature* tell the story: "An Unstable Intermediate Carrying Information from Genes to Ribosomes for Protein Synthesis" by Sydney Brenner, François Jacob, and Matthew Meselson and then "Unstable Ribonucleic Acid Revealed by Pulse Labelling of Escherichia Coli" by another group led by James Watson and Walter Gilbert at Harvard and colleagues at the Institute Pasteur in Paris.[9,10] The implication of this work was that mRNA molecules shuttled information between DNA in the nucleus to ribosomes in the cytoplasm. Thus, a copy of a gene was made by constructing an RNA sequence, a process later known in molecular biology parlance as gene transcription—the writing of gene sequence (DNA) into RNA.

Determining the genetic code would take years of work with the contribution of many scientists to finally reveal nature's solution. Sydney Brenner, working with Francis Crick at the Cavendish lab in Cambridge, put forward ideas in a manuscript that laid out the first problem to be experimentally addressed. How could any such code, using only four letters (DNA's A-adenine, C-cytosine, G-guanine, and T-Thymine) specify which one of 20 amino acids was to be incorporated into a protein at a particular location in the molecule? A code using a single letter to represent a set of 20 amino acids would be insufficient as would taking two letters at a time (in other words, AT, AC, GG, and so forth), which would allow encoding of only 16 (4^2). That leaves a minimum of a three-letter code (a triplet such as CCA) that would provide for up to 64 (4^3), or 4 (4^4 =256), and so on. Any coding system using three or four nucleotides at a time would have redundancy, with more than one triplet or quadruplet allowed to code for one amino acid (for example, CCA, CCG, and CCC used for proline). Would nature use a degenerate code, and what would the code look like?

In a culmination of the quest led by Francis Crick at the Cavendish Laboratory in Cambridge, he and Leslie Barnett, Sydney Brenner, and Richard Watts-Tobin laid out the logic for the solution by combining their work with others in a paper in *Nature* at the end of 1961 entitled "General Nature of the Genetic Code for Proteins," unlocking the mystery.[11] From their genetic experiments using "frameshift" mutations, they deduced correctly that the genetic code utilizes triplets and therefore must harbor degeneracy. The code was determined also to use nonoverlapping sets of three letters, known as *codons*, to specify individual amino acids. Thus, amino acids could be coded for by more than one codon. At the time, only one codon had been determined—that for the amino acid phenylalanine. Years would stretch on as imaginative chemistry by Gorbind Khorana and structural mapping of tRNAs (including the very first sequencing of a nucleic acid for the 77-nucleotide tRNA encoding alanine in Robert Holley's lab) finally filled out the details of the 64-entry, 3-letter word dictionary of the genetic code and the elaborate orchestration of protein synthesis. These fruitful endeavors were acknowledged with the awarding of a Nobel prize in 1968 "for their interpretation of the genetic code and its function in protein synthesis."

The fundamental discoveries of the structure of DNA, the process of information flow whereby DNA makes RNA makes protein (in other words, the central dogma)—as directed by RNA transcription and protein translation—had been worked out by the mid-1960s. The genetic code was universal: it did not change across organisms. It would take 20 more years before molecular techniques based on enzymatic, fluorescent-based sequencing and instrumentation developed in Leroy Hood's laboratory at the California Institute of Technology in Pasadena would enable sequencing of small segments of DNA, 100 nucleotides in length. It was hoped that technologies would improve to enable rapid sequencing of hundreds of samples a day and genes in the range of

1,000 to 10,000 nucleotides. Prior to the introduction of computing and information technologies, it was unfathomable that genome-size pieces such as human chromosome 1, which possesses 250 million base pairs, could be analyzed and decoded. What were the prospects of learning anything from genome sequencing if out of the entire book of life, one could read only a few sentences, and those were taken at random? The incredible vault of life's information stored in DNA was still remarkably inaccessible well up until the 1990s.

Manipulating Gene Information with Recombinant DNA Technology

The astounding successes in defining the genetic chemistry of prokaryotes during the 1950s and 1960s were both exhilarating and challenging. Not surprisingly, I and others wondered whether the more complex genetic structures of eukaryote organisms, particularly those of mammalian and human cells, were organized and functioned in analogous ways. Specifically, did the requirements of cellular differentiation and intercellular communication, distinctive characteristics of multicellular organisms, require new modes of genome structure, organisation, function and regulation? Were there just variations of the prokaryote theme or wholly new principles waiting to be discovered in explorations of the genetic chemistry of higher organisms?

—Paul Berg, Nobel Lecture, 1980

The remarkable successes of bacterial genetics pushed molecular biology deeper into new terrain. Paul Berg at Stanford University and a new crop of molecular biologists wanted to go beyond bacterial systems and bacteriophages (viruses of bacteria) to probe the fundamental nature of genome structure, gene regulation, and function in higher organisms. With a confluence of fortunate discoveries at their disposal, these "gene jockeys" could begin to address the larger questions by using newly developed molecular tools and their imaginations to manipulate biological systems. There was a sense of anticipation that the experimental possibilities with gene manipulation power were nearly unlimited.

Ultimately, the first goal had to be the development of a new technology framework, allowing experimenters to isolate and amplify genes and other DNA segments, edit their sequences, combine them, and produce copies. Since all of the progress was in prokaryotic systems, the work would naturally start from there. They already had in hand functioning DNA elements from bacterial genes, enzymes that catalyzed reactions with DNA as a substrate, and viruses and plasmids to engineer new creations. The ultimate aim was to be able to extract genes from higher organisms, reconfigure them, and then place them back into their natural context. Transgenes and transgenic animals were built in order to examine a gene's effects on normal cellular processes and, at some point, to introduce genetic changes into model systems of disease to understand a gene's role in disease.

Since the turn of the 20th century, modern biology had advanced by descriptive studies with limited ability to perturb cell function. Chemists studied macromolecular synthesis and dissected metabolic pathways in test tubes, not living cells. Geneticists worked out laws of heredity, recombination, and gene organization by performing relatively slow genetic crosses in their favored organisms (corn, fruit flies, bread molds) and then measuring phenotypic traits in offspring. The only gene variants that could be studied were those found naturally occurring in a population (for instance, Mendel's smooth or wrinkled peas; Thomas Hunt Morgan's flies with crumpled wings and white eyes). Could molecular biology turn this around and deliver technology to engineer genes to encode specific functions directly into organisms?

For the molecular biologists, the logical first step was to create artificial hybrid DNA molecules: one part from bacteria and one part from a different virus or organism. This concept had long been on Berg's mind, and his lab was on its way to becoming the first to engineer a so-called recombinant DNA. After working with Renato Dulbecco at the Salk Institute on the molecular biology of SV40, a DNA virus with oncogenic potential, Berg envisioned being able to use viral strategies to move foreign genes into mammalian cells. The recombinant DNA molecule would be comprised of a lambda phage DNA segment containing the bacterial galactose operon, together with SV40, which had a circular genome. To open up the circle, the DNA of SV40 would need to be cut and then the lambda DNA could be joined at both ends and re-circularized. In the spring of 1971, the Berg lab was preparing to make the hybrid molecule, but fate, and an abundance of caution, intervened to hit the brakes on recombinant DNA technology for years to come.

The nascent field was moving quickly to create molecules never seen in nature; the step in this direction was similar to the biological chemists of an earlier era that created non-naturally occurring chemotherapy drugs based on rational design. Herb Boyer and Stanley Cohen at Stanford were working together to utilize bacterial plasmids to introduce foreign DNA into bacteria, a practical alternative. Although Berg was interested in using SV40 as the delivery vehicle into bacteria and not mammalian cells, there were still concerns that this virus in a new hybrid molecule might be harmful and inadvertently become a biological pathogen. Berg's graduate student, Janet Mertz, had traveled to Cold Spring Harbor for a 1971 summer conference and revealed the nature of the laboratories' work. Learning of the work for the first time, scientists expressed concerns as to why they were using an oncovirus in a potentially uncontrolled way in very early experiments. One of James Watson's recruits to Cold Spring Harbor and staff scientist, Robert Pollack, raised the alarm about SV40 and proceeded to call Berg to voice his strong objection.

Soon, Berg was on calls with others and arranged a meeting to discuss all of the potential issues, keeping the discussions private among a small group of scientists in the United States. A moratorium on recombinant technology using SV40 and other DNA of unknown hazardous potential was suggested to investigate safety issues further. In late 1971, Berg voluntarily put an indefinite hold on the next step of his team's recombinant DNA research, which was to proceed with the gene transfer step into *E. coli* and evaluate whether the artificial DNA would survive in a bacterial host and also be replicated.

These sorts of concerns would explode on the world stage over the next few years, with Paul Berg playing a major role in navigating the torturous private and public debates, setting recommendations that followed from scientific conferences convened to develop guidelines. The drama culminated at the International Conference on Recombinant DNA Molecules at Asilomar Conference Center in Pacific Grove, California, over three days in late February 1975, that was organized and chaired by Berg.[12]

Although Berg halted his work due to the connection of SV40 to cancer, other laboratories around the world continued to press forward with bacterial virus and plasmid DNA manipulation. Forty miles north of Stanford at the University of California, San Francisco (UCSF), Herb Boyer's lab happened to be doing similar experiments with bacterial DNA and had purified a critical tool; it was the bacterial enzyme known as *Eco*RI. This workhorse of molecular biology is one of the molecular scissors that cuts DNA at a hexameric site. *Eco*RI recognizes the palindromic sequence GAATTC in double-stranded DNA. Although hundreds of these types of restriction endonucleases can be purchased from catalogs today, researchers then relied on gifts from a small handful of scientists with the technical skills to characterize and purify sufficient quantities for experimentation. The enzyme was shared with laboratories at Stanford, and to Berg's delight, this tool was perfect for his experiment, as *Eco*RI would only cut SV40 once, opening up and linearizing the virus genome. As soon as the Berg laboratory had completed and published the first gene-splicing experiment with SV40 and bacterial DNA in 1972, with the help of Boyer's *Eco*RI enzyme, recombinant DNA technology was born.[13]

Boyer had formed a collaboration with Stanley Cohen at Stanford to build a plasmid-based shuttle system. Similar to Berg, they wanted to develop a bacterial system to provide a way to retain and replicate artificial hybrid, or recombinant, molecules. Cohen was one of the first to exploit the use of bacterial plasmids that carry antibiotic resistance genes (the first plasmid he used was pSC101, planting his initials into recombinant DNA lore).[14] His earlier work had figured out a chemical process for promoting the transfer of plasmid DNA into *E. coli* bacteria, which was termed *transformation*.[15]

The two San Francisco Bay Area laboratories were about to test another recombinant DNA idea: Could the antibiotic resistance plasmids be used as an identifier of those bacteria that had acquired the new recombinant molecules? The initial design was to use two antibiotic resistance genes, for tetracycline and kanamycin, and the nonpathogenic *E. coli* strain C600. If plasmids harboring one or both of these genes were functional, then the bacteria could survive on surfaces (agar plates) containing these antibiotics. Using the tools of molecular biology to cut and paste together the artificial sequences into the plasmid, including a hybrid with both plasmids mixed together, they demonstrated the feasibility of the recombinant DNA process by growing antibiotic resistant bacteria on either kanamycin or tetracycline agar plates. Additional experiments proved that the recombinant DNA harbored biologically functional molecules. The method had obvious practical applications: first, once bacterial colonies appeared on antibiotic-containing plates, the bacteria could be cultured and grown in large quantities and plasmid DNA recovered; second, if a molecular cloning experiment worked, you knew immediately—a bacterial colony with antibiotic resistance meant that your foreign piece successfully ligated into the plasmid. As for pSC101, the authors concluded the following:

> "The antibiotic resistance plasmid pSC101 constitutes a replicon of considerable potential usefulness for the selection of such constructed molecules, since its replication machinery and its tetracycline resistance gene are left intact after cleavage by the EcoRI endonuclease."[15]

The important results were scheduled to be published only in July, just after the Gordon Conference on Nucleic Acids, which began in New Hampton, New Hampshire, in June 1973. However, at one of the most closely watched sessions of the meeting, Boyer was scheduled to speak. Against the wishes of his UCSF collaborator Stan Cohen, Boyer had let slip that cloning of the *Eco*RI fragments of two different plasmids into a recombinant molecule was a success and that their transformation of *E. coli* had worked, providing the bacteria with newly acquired antibiotic resistance. A shock wave ensued. It was now clear that, given a few simple tools, researchers could take foreign DNA from anywhere, practically, and propagate non-natural DNA sequences in bacteria. Now it was just a matter of putting in DNA segments of interest, including viral sequences and nonbacterial genes from frogs, mice, and even humans.

Almost immediately, a group of scientists convened at the end of the conference to grapple with the potent consequences of allowing the technology to proceed, unguided. A letter was sent to the presidents of the US National Academy of Sciences and the National Institute of Medicine. A few of the significant portions follow:

> *"We are writing to you, on behalf of a number of scientists, to communicate a matter of deep concern. Several of the scientific reports . . . indicated that we presently have the technical ability to join together, covalently, DNA molecules from diverse sources . . . This technique could be used, for example, to combine DNA from animal viruses with bacterial DNA, or DNAs of different viral origin might be so joined. In this way new kinds of hybrid plasmids or viruses, with biological activity of unpredictable nature, may eventually be created. These experiments offer exciting and interesting potential both for advancing knowledge of fundamental biological processes and for alleviation of human health problems.*

> *"Certain such hybrid molecules may prove hazardous to laboratory workers and to the public. Although no hazard has yet been established, prudence suggests that the potential hazard be seriously considered."*

The letter was published in the journal *Science* on September 21, 1973, and it was the first public warning shot.[16] Worded carefully, the authors, Maxine Singer and Dieter Söll, shied away from mentioning a major source of anxiety, the potential of these methods to unleash cancer.

In retrospect, it is quite easy to see the scientific and cultural forces at play that led to such a pronouncement and the desire to proceed with caution. At one level, the most valid concern was that an engineered plasmid introduced into bacteria might be carrying around an oncogene or other genetic control segments that could cause uncontrolled growth or lead to a novel, pathogenic bacterial strain. For example, *E. coli* strains exist in the human gut and cause no harm. Some, such as

E. coli O157, cause serious gastroenteritis. However, a laboratory-created strain might inadvertently wind its way into the human gut, alter the microbiome, and transfer genes to other *E. coli* species by a process known as *horizontal gene transfer*. Worse, some pondered whether recombinant DNA from rogue bacteria could be transferred into human intestinal cells and cause cancer.

Most of the researchers developing and utilizing molecular biology tools were virologists, and the consensus of this scientific community was that viruses caused cancer. Since the 1950s, there were three competing hypotheses about what causes cancer, and these were thought to be mutually exclusive. There were the virologists who claimed that viruses cause cancer; epidemiologists who believed that exogenous substances (chemicals mainly) were causal and triggered uncontrolled cellular growth; and a few that considered alteration of genomic elements as the underlying cause. Evidence was flimsy at best for all three theories. A spark of evidence had shown up in 1960 when SV40, the monkey virus, was found to cause tumors in a rodent model. SV40 thus became the focus of research programs aimed at understanding how viruses cause cancer throughout much of the 1960s.

In 1959, an earthquake of an experiment was performed by Howard Temin with Rous sarcoma virus that shook the cancer field. He showed that he could create "cancer in a dish" by infecting cells with the virus that has an RNA genome. The genetic material of the virus appeared to have been integrated with the DNA of the cells. The result cut completely against the grain of the central dogma. Temin suggested that information flow might be possible in reverse, RNA copied to DNA. A decade later in 1970, he and David Baltimore independently discovered reverse transcriptase, named so because the enzyme could take RNA as template to make a DNA product, the "reverse" of well-established dogma. At the same time, virologists, notably Peter Vogt, Peter Duesberg, and Steve Martin, were using genetic and biochemical methods to home in on the gene conferring oncogenic transforming activity. Using transformation-defective mutant viruses, they identified a single gene they called *src*, the first oncogene. The molecular mechanisms underlying human carcinogenesis seemed well within reach through the lens of these virologists.

As the virologists were chasing down evidence for additional cancer-causing viruses and their DNA fingerprints linking genes to cancer, Berg and others were attempting to exploit the properties of these same viruses to build a new technology and someday advance these tools into medical practice by moving into eukaryotic systems. The early recombinant DNA methodologies and concepts developed by these pioneers led directly to ideas for their potential uses in gene therapy, as Berg foresaw so early on.

The period following the Gordon Conference from June 1973 all the way up to Asilomar in early 1975 saw many recommendations and more controversy on how best to accommodate experimentation and assuage public safety concerns. During this time, Cohen and Boyer's collaborative effort dropped another bombshell. They published the long-sought technical achievement of combining DNA from a more complex organism, in this case two genes from the African clawed frog *Xenopus laevis*, into bacteria using the Cohen pSC plasmids.[17] The stage was now set for compelling all of the players to take action on the future of molecular cloning technology. The "MIT meeting" in April 1974 delivered four recommendations on commencing experiments; the Asilomar meeting in February 1975, with worldwide press coverage, brought consensus on a safe path forward.

Following an intense three days of discussions, Berg, together with many other molecular biologists and future Nobel laureates, put in place regulations on recombinant DNA technology. The guidelines entailed that laboratories develop biocontainment facilities, including physical and biological barriers, to prevent contamination and spread of potential pathogens. The ability of scientists to achieve this on their own, without it disappearing into murky legislation or potential bans, was a landmark achievement. In doing so, recombinant DNA gained public acceptance as a breakthrough technology with anticipated benefit for humanity. The meeting at Asilomar is now regarded as a paradigm for how to approach fundamental technological advances that can cause permanent changes to society, alter human life, and impact the biosphere. I will return to societal issues in later chapters as exponential progress in artificial intelligence and the latest genetic engineering from the 21st century will lead industry to another technological horizon (that challenges conceptions of life).

Genetics, Gene Discovery, and Drugs for Rare Human Diseases

Molecular biology continued to expand the recombinant DNA toolkit with technical advances to analyze nucleic acid sequences in the late 1970s. The pioneers of the molecular biology era were at work developing two distinct approaches to "read" a DNA sequence. Fred Sanger, who had earlier won a Nobel prize in Chemistry for techniques to sequence proteins, developed what was ultimately to become the mainstay procedure. His ingenious approach was to use chain terminating inhibitors of each of the four DNA bases, such that DNA strands could be synthesized by DNA polymerase reading off of its template until it incorporated a nucleotide with a dideoxy base, which prevented chemical linkage of the growing DNA copy. Images in the popular media today of X-ray films contain the archetypal DNA banding pattern are products of Sanger reactions (and iconography of antiquated technology). Allan Maxam and Walter Gilbert developed an alternative DNA sequencing method based on chemical modification of DNA in one step and subsequent cleavage at specific bases in a second step. Maxam-Gilbert sequencing had an initial accuracy edge over the Sanger method and the advantage of allowing purified samples of DNA to be used without further cloning. However, this was a technically difficult method and laboratories eventually favored Sanger sequencing after refinements to dideoxy sequencing had been implemented in the early 1980s. The development of DNA sequencing technology was a gateway to retrieving and recording the information in genomes. What remained was the automation of these arduous laboratory processes to enable widespread use for basic research, diagnostics, and therapeutic development.

Although decades away from gene-based therapies, advocates for rare diseases together with geneticists were about to launch monumental disease gene discovery projects that could shed light on paths to new treatments. The mapping of disease genes began in parallel with DNA sequencing technology and utilized the tools of genetics—linkage analysis and fine mapping in large, well-characterized pedigrees. By tracking the inheritance of DNA markers that traveled along with affected individuals of a particular disease, scientists could begin the hunt for a chromosomal location and eventually determine the DNA sequence of the disease-causing allele. Establishing that the underlying genetic variants were causal required evidence that the risk genotype was absent in healthy members among unrelated pedigrees with a familial history of disease. More laborious work was needed to determine the functional consequences of the genetic mutations, relying on animal models and biochemistry that tended to grind on for years of investigation to reach an answer—or remain a mystery. Huntington's disease exemplified the latter.

Nancy Wexler at the Hereditary Disease Foundation and Columbia University in New York led a large, decade-long collaborative effort that culminated in the discovery of the Huntington's disease gene on chromosome 4. The journey began with Wexler's heroic efforts to obtain pedigree information and blood samples for DNA marker analysis from an extensive set of families in Venezuela living around Lake Maracaibo, all of whom descended from a woman with the disease living in the 19th century. Figure 4.3 pictures Nancy Wexler holding a large pedigree chart filled with data and symbols representing affected individuals (black circles) and other family members. More than 10,000 DNA samples were used for Huntington's research from this community over the years.

Figure 4.3: Nancy Wexler with the Venezuelan family pedigree chart for tracking inheritance of Huntington's disease
Dr. Nancy Wexler, unrolling a section of a wall-sized genetic chart of families she has researched for Huntington's disease, Columbia Presbyterian Hospital.
(Photo by Acey Harper/The LIFE Images Collection via Getty Images/Getty Images). Image licensed for use by author.

Wexler, together with James Gusella at the Massachusetts General Hospital, had built a strategy based on a meeting in 1979 where they discussed using Botstein's RFLP-based genetic methods. By 1983, the team had used RLFP techniques and determined that the Huntington's gene was located near the tip of the short arm of chromosome 4. This was the first mapping of a disease gene to a genomic position using genetic markers and patterns of inheritance. The genetic abnormality was caused by a trinucleotide repeat in the gene's protein coding sequence (CAG repeats of the glutamine codon), which appeared as variable length expansions. The protein sequence offered no clues as to its biological function, but a growing number of other neurological disorders featured a similar unstable genomic element as the culprit. For decades, attempts to understand the function of the huntingtin protein and what drove the pathology in the brain came up empty.

The landmark discovery and completion of the sequencing of the Huntington's disease gene[18] in 1993 marked the beginning of a torrent of successes in identifying genes underlying rare, Mendelian disorders, cancer, and mitochondrial diseases, as summarized in Figure 4.4. The use of recombinant DNA tools, PCR, genome sequencing, comparative genomics, and genetic studies led to the uncovering of mutations causing 1,000 single gene disorders by the close of the 20th century and more than 5,000 by 2020. What is left to be found? Estimates vary, but there are probably at least 7,000 monogenic disorders, 80 percent of which are genetically based. For those remaining with an undetermined cause, these are almost certainly going to be ultrarare disorders affecting single individuals or a few families.

For Huntington's disease, much more is now known about the cellular processes driving the neurodegenerative pathology, but the quest for a therapeutic continues down a long road, decades past the discovery of the mutation in the gene. Rare diseases like Huntington's share a difficult path for finding a cure for several reasons. Research linking the genetic mutation to a protein's functional consequences is not straightforward, as exemplified by huntingtin's enigmatic role in normal physiology. For drug developers, rare disease may not present an identifiable target, or the medicinal chemists might be handed a protein (like huntingtin) that is not druggable by conventional approaches. In many rare diseases, translational scientists lack robust animal models that faithfully replicate the human illness. From a pharmaceutical business model standpoint, market considerations are paramount. Developing a product around a poorly understood medical condition and for its small, rare disease population might only be justified if the cost of the therapy delivered per patient was exorbitantly high. Such concerns exemplify the odyssey of rare disease therapeutics and the exasperation felt by patients and caregivers.

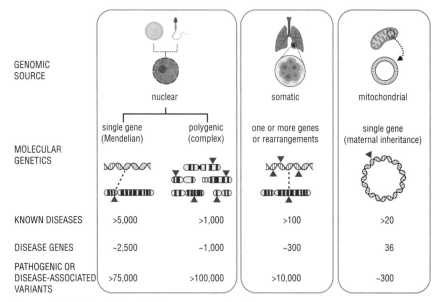

Figure 4.4: Molecular basis of human genetic disease

DNA alterations underlying human disease arise from three genomic sources with distinct molecular genetic features.

Left panel: Nuclear DNA harbors all of the variation acquired from the germ cells (sperm and egg) of the parents, where pathogenic variants are referred to as *germline mutations*. Single gene disorders follow a Mendelian pattern of inheritance where gene defects such as single base changes, insertions, or deletions that affect the encoded protein's function or expression cause disease when present in one (dominant) or both alleles (recessive) for genes on autosomes (nonsex chromosomes). The ClinVar database has cataloged 75,000 pathogenic variants across the human genome (as of August 2020).

Middle panel: DNA mutations arising in somatic cells can lead to cancer as the result of driver mutations in a small set of genes. Somatic mutations can combine with germline mutations in oncogenes or tumor suppressors to predispose individuals to certain types of cancer.[19] There are approximately 100 general types of cancer and more than 500 rare cancer subtypes.

Right panel: The mitochondrial genome is acquired solely by maternal inheritance. The small, circular genome (~16,000 nucleotides) encodes 37 mitochondrial genes; mutations in 36 of these genes at 300 known pathogenic sites cause 20 mitochondrial disorders. There are many mitochondrial diseases that arise due to mutations in nuclear-encoded mitochondrial proteins (more than 1,100 proteins are imported to the mitochondria to support this organelle's function).[20]

For the vast majority of rare diseases, there is no treatment, with medicines available for only about 500 of the many thousands of known diseases. The exception is in cancer, where rare subtypes have driven research and development efforts leading to many approved drugs (see Chapter 3, "The Long Road to New Medicines"). Advocacy from organizations such as Wexler's Hereditary Disease Foundation, the National Organization of Rare Diseases, and the public awareness brought about through President Nixon's war on cancer in the 1970s (the National Cancer Act of 1971) led to a change of fortune for rare disease research and shaped drug development for decades. At the inception of the biotechnology era, steps were taken by regulatory agencies and Congress to incentivize universities and the drug industry for taking risks in research and development and explicitly for taking on projects for rare diseases. The Bayh-Dole Act of 1980 allowed universities to patent and retain ownership of inventions arising from federally funded research at their institutions. Thus, for almost any disease-related research, there was a path to commercialization for potential products, or technologies, through licensing and technology transfer. In retrospect, it is clear that this greatly accelerated and enabled the blossoming of the biotechnology industry in the United States. The second key part was the Orphan Drug Act, which was legislation introduced to the House of Representatives by Henry Waxman in 1982 and signed into law by President Reagan in 1983. The law designated orphan drug status to any product where the prevalence of disease in the United States was 200,000 or fewer individuals (roughly 1 in 2,000 people). As a group, rare diseases affect 30 to 40 million in the United States and a similar number in Europe. With legislative changes enacted to impact these people's lives, the industry was poised for change.

Were the right incentives put into place for industry to search for cures instead of blockbuster drugs? Companies received tax credits, fee waivers, and a seven-year market exclusivity for any orphan drug approval. The latter turned out to be the game-changer. By almost any account, the regulatory changes dramatically altered the therapeutic development landscape. Over the time period from 1983 to 2017, drugs with an orphan designation from the FDA accounted for more than 25 percent of all approvals for NMEs. The industry's shift toward biologics since 2000 also includes those with orphan designation, where orphan drugs have captured 40 percent of the biological approvals. The caveat here is that oncology drug development dominates the orphan category, accounting for more than half of the clinical trials initiated in the past several years.

The arrival of precision genome engineering tools will likely allow rare disease indications a chance to be evaluated in research and development pipelines and play catch-up. For genetic diseases, the most prevalent type of pathogenic variant is known as a *point mutation*, or a *single nucleotide variant (SNV)*. Table 4.1 lists variant classes in genetic disease and their relative abundances, as found in the ClinVar database.

Table 4.1: Types of Pathogenic Variants Underlying Human Genetic Diseases

GENETIC SCALE (BASE PAIRS)	VARIANT TYPE	RELATIVE ABUNDANCE	REPRESENTATIVE DISEASE
1 base pair	Point mutations (single nucleotide variants, SNVs)	60%	Sickle Cell
10^0–10^2	Micro deletions and insertions	25% and 5%	Tay-Sachs
10^3–10^4	Copy number variants	5%	Prader-Willi / Angelman Syndromes
10^4–10^7	Large-scale rearrangements	<5%	Charcot-Marie Tooth
10^7–10^8	Chromosomal loss/ duplication	<1%	Down Syndrome

The term SNV covers the spectrum of annotated variants in four classes that are described as clinically benign (no effect on protein function or phenotype), a variant of unknown significance, likely pathogenic, and known pathogenic.

For single-gene diseases, point mutations tend to disrupt the coding region of a protein sequence and may reside at one unique position. Alternatively, the same disease may be caused by any one of thousands of variants found throughout the extent of the gene. Conventional gene therapy approaches are not capable of editing or changing a genomic mutation and instead seek to supply a full version of a corrected gene to replace or augment the disease allele. The gene's instructions are sent to the cells or tissues via a nonreplicating viral vector to get recipient cells to manufacture the needed protein, correct the dose, or counteract the deleterious gene's product in some way.

Tackling the genetic defects associated with polygenic disorders, such as diabetes, hypertension, or a number of more common psychiatric diseases, remains problematic. Genetic studies have uncovered the contribution of many risk alleles for common conditions, but thus far have not yielded clear therapeutic opportunities at the variant level. Most of these disease-associated variants are SNVs and lie in noncoding, potentially regulatory regions of the genome. Altering or tuning gene regulatory circuits for therapeutic outcomes will require combinatorial strategies. Turning genes on or off for monogenic diseases has attracted attention due to the underlying genetics (for example, shutting off a damaging allele in an autosomal dominant disorder such as Huntington's disease) and ease of modifying a single regulatory site. Addressing all of the pathogenic variant types for monogenic diseases, once considered fantasy, has now entered the realm of possibility with genome editing by use of programmable nucleases.

Second-Generation Biotechnology Tools: CRISPR-Cas9 and Genome Editing Technologies

The simplest living organisms, the prokaryotes, together with their quasi-living predators, coevolved molecular surveillance and interference systems more than a billion years ago and have now made their way into the laboratories of biologists. For the prokaryotes, it was vital to deal with external threats and parasites, such as bacteriophages and other invasive mobile genetic elements (plasmids and transposons) in their hazardous environments. Starting around 2002, microbiologists recognized that bacterial and archaeal genomes contained a family of repeat DNA sequences that were recognized as "clustered regularly interspaced short palindromic repeats" and referred to them as CRISPR.[21] In 2005, several research groups discovered that the spacer DNAs found in these CRISPR arrays were essentially copies of viral and plasmid sequences.[22–24] This observation was one of the first clues that perhaps these single-cell organisms were storing and then somehow using snippets of invader DNA for a search and destroy capability. The core elements of the defense system came into view once CRISPR-associated

(Cas) proteins and CRISPR RNAs (crRNAs) were uncovered at the same genomic locus. Genetic experiments proved that CRISPR arrays and Cas proteins were critical for protection against phages.[25] The *Streptococcus pyogenes* CRISPR-Cas9 type II system performed its rapid destruction of bacteriophage DNA by employing a Cas nuclease (Cas9) and two distinct noncoding crRNAs, a transactivating crRNA (tracrRNA) and a nuclease guide sequence crRNA. As soon as Doudna and Charpentier's publication appeared in *Science* in 2012, demonstrating that a two-component system (composed of Cas9 and a single fused RNA guide) could be built to edit any arbitrary DNA sequence, a powerful new biotechnology was born.

The immediate impact of the simplicity built into the design of the new CRISPR-Cas9 system was that researchers could begin experiments in mammalian cells and engineering could commence to build an array of versatile tools. In the first CRISPR-Cas9 system that was proposed, a single guide RNA acted to "program" the Cas9 endonuclease. The RNA was a sequence-specific targeting mechanism and Cas9 cleaved DNA to leave blunt ends that could be repaired by cellular DNA repair mechanisms, or with directed strategies, as outlined in Figure 4.5. Other labs raced to develop new systems and envisioned new applications for research biology and regenerative medicine. Feng Zhang's team at the Broad Institute was the first to re-engineer the Cas9 protein to enable high-fidelity repair and demonstrate a functional CRISPR-Cas system in operation in mouse and human cells.[26]

CRISPR systems were the second wave of technologies that had been developed to perform gene editing with programmable nucleases. The earlier methods were based on protein engineering of DNA modifying enzymes to produce meganucleases, zinc-finger nucleases (ZFNs), and transcription activator-like effector nucleases (TALENs). The programmable feature of these systems rested in designing a DNA-protein binding site into the enzyme by fusing a DNA recognition motif together with an endonuclease activity. The protein engineering and optimization steps for ZFNs and TALENs were often difficult procedures that were done commercially by a few biotech companies and offered to researchers at considerable cost. The ease of manipulating CRISPR systems and the startling array of prokaryotic defense systems with which to exploit—now more than 30 subtypes are known with more to be discovered—has turned the entire genome engineering field into a biotechnology juggernaut.

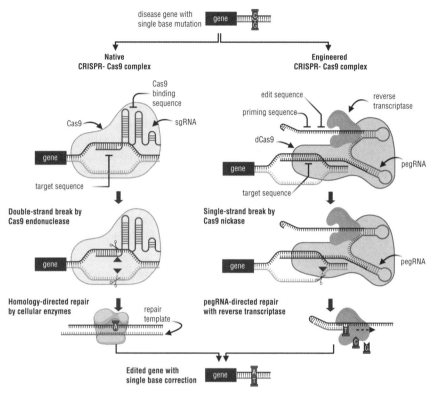

Figure 4.5: Genome editing strategies with programmable Cas nucleases and template-directed repair

The development of genome editing technologies began with a modification and fusion of the RNAs needed to guide target DNA into a native CRISPR-Cas9 complex (left side of the flow diagram). The elements of a programmable nuclease include the Cas9 endonuclease and a single RNA guide (sgRNA) molecule containing a Cas9 binding sequence and a 20-nucleotide target sequence complementary to a gene region of interest. Binding of the sgRNA and conformational changes in Cas9 activate the enzyme's endonuclease activity at a site near the protospacer adjacent motif (PAM sequence not shown) to produce a double-strand break in the gene. The second element of the technology is to co-opt cellular DNA repair mechanisms to recognize the double-strand break and direct the enzymes to use a repair template to perform the "edit." In this example, the disease form of the gene has a C:G base pair (top of diagram), and the editing corrects this to leave an intact gene with a A:T at the same position (bottom of diagram). This type of edit (a transversion) cannot be performed by current base editors. Genetic and protein engineering has led to the development of Cas9 lacking endonuclease activity (dCas9) but built with other functional components to perform editing without the need to rely on cellular DNA repair processes. A prime editor (right side of the diagram) uses a reconfigured guide RNA that contains a priming and edit sequence, along with a gene targeting sequence (pegRNA). The engineered Cas9 possesses a nickase that creates a single strand break on the target gene; a reverse transcriptase performs the edit by synthesizing a new strand based on the edit sequence.

GENE EDITING TOOLS BUILT ON ENGINEERED CRISPR-CAS SYSTEMS

Target disruptors: Programmable Cas nucleases such as natural Cas9 disrupt a genomic locus by producing indels within the target sequence. (CRISPR Therapeutic's first gene editing target is a regulatory site disrupted with this tool). Cas9 uses the guanine-rich PAM sequence 5'-NGG-3', which is required to be adjacent to the target sequence. Cas12a variants have thymine-rich PAM sequences for targets that are A/T rich. Cas12a and Cas14a are also unique in their ability to nonspecifically cut single stranded DNA, which has been exploited for CRISPR-based diagnostics. Several Cas9 variants have been engineered to decrease off-target editing and improve fidelity of on-target edits (sniper-Cas9, hypaCas9, SpCas9-HF1, and eSpCas9) or extend targeting capability (xCas9). A third system based on protein engineered Cas12b is also suitable for human genome editing.

DNA base editors: A class of editing technologies to perform single base editing of transition point mutations. These are built on a dCas9 chassis that does not introduce double-strand breaks in DNA and has Cas nickase activity instead for single-strand cuts. Cytosine and adenine deaminases are engineered into dCas9 to achieve base-specific edits, along with other components to improve editing efficiency. At least a dozen types have been engineered or created by directed evolution. These editors are further categorized by the size of the editing window across the target and by sequence context preference (currently only for cytosine base editors).

RNA base editors: A base editing technology developed to install edits in RNA. Two systems have been engineered, one using dCas13 fused together with ADAR2 or with ADAR2 evolved into a cytidine deaminase, named RESCUE. A second is based on the RNA-targeting Cas9, or RCas9.

Prime editors: Gene editor with full range of point mutation editing (transitions and transversions) and small indel creation capabilities. Prime editors are engineered by fusing a Cas nickase with reverse transcriptase and requires a guide RNA with several components (see Figure 4.5).

Target integrators: CRISPR-associated transposases and recombinases could provide the ability to insert or move large DNA segments into genomes into specific sites. The emerging tools combine Cas proteins and transposon-associated components. Zinc finger nucleases have previously been fused with site-specific recombinase catalytic domains; CRISPR-Cas recombinase fusions are in progress.

Human Genome Editing and Clinical Trials

Medicine is entering an era where CRISPR-Cas tools for genome engineering will be used therapeutically to address genetic diseases resulting from rare, monogenic conditions and improve engineered cellular therapies for cancer.* How many diseases can be cured by gene editing and the number of disease-modifying therapeutics that will be widely available for addressing the needs of the global population are unknown. Data from published pipelines, preclinical studies, and ongoing clinical trials from gene therapy-focused biotechnology and pharmaceutical companies indicates that there are approximately 100 therapeutics under clinical development targeting close to 50 diseases. While gene editing technology is deceptively simple in operation, its application in medicine involves numerous complexities around safety, delivery, durability, efficiency, manufacturing, and cost (to patients) — all of which indicate a long implementation phase comparable in scope to that experienced during the development of gene therapy 1.0.

On issues of safety, two elements stand out. First, similar to gene therapy in the clinic, the immune reactions to viral proteins, and potentially the CRISPR-Cas components themselves, remain a chief concern. Much of the field has shifted to the use of less immunogenic adeno-associated viruses (AAVs) or lentiviruses for therapies requiring *in vivo* administration. The ability to deliver genetic material in lipid nanoparticles (LNPs) is also gaining significant traction, particularly with the safety profile and success of the LNP-encapsulated mRNAs that were proven in clinical trials of Moderna Therapeutics' and Pfizer-BioNTech's vaccines against SARS-CoV-2. The second safety issue concerns the engineering precision of the programmable nuclease editors—all current systems have a measurable rate of aberrant gene editing at the desired site and some are known to generate unpredictable, off-target edits that could cause deleterious mutations in important regions of the genome or even alter entire chromosomes. Preliminary studies that evaluated CRISPR-Cas editing in early human embryos reportedly found frequent loss of heterozygosity, implying that large genomic segments or entire chromosomes were lost in dividing cells.[27] Although human embryo editing is clearly unwarranted for the vast majority of diseases and unethical at this stage,

---

*Mitochondrial DNA repair with CRISPR-Cas systems is not yet feasible due to difficulties in delivering guide RNAs within the matrix compartment where the genome resides, which is buried behind the organelle's double membrane.

the possibility of such catastrophic changes in the genome of developing human tissues ensures that genome editors will be tested extensively in model systems and human cells prior to initiation of interventional studies.

As these and other concerns have arisen, a set of next-generation sequencing technologies have been developed to detect off-target events, namely, GUIDE-Seq and CIRCLE-Seq. The off-target characteristics of CRISPR are due to target specificity of the single guide RNAs. Extensive screening and selection of highly specific guide RNAs can circumvent this aspect of precision. Strategies for implementing therapeutic gene editing share delivery methods in common with gene therapy, employing either *ex vivo* or *in vivo* routes of administration. Figure 4.6 shows an overview of the conceptual approaches and steps involved in therapeutic gene editing.

Over the past several years, the native bacterial CRISPR-Cas systems have been engineered not only to widen the range of applications and editing targets (to include all nucleotide substitutions, targeted deletions, or insertions) but also to improve the precision of gene targeting, editing fidelity and efficiency, all anticipating its eventual clinical use. The development of DNA base editors and their testing in mammalian cells illustrates the road ahead. Cytosine and adenine base editors initially appeared promising to execute single base changes on selected DNA targets and possibly would be favored as tools in the clinical use of gene editing. But as confidence was building, researchers showed that some base editors engaged in highly promiscuous editing of RNA transcripts—with cytosine base editors inducing cytosine to uracil changes across tens of thousands of transcripts in many genes. A similar off-target RNA editing ability was found with adenine base editors. The results were a surprising revelation: a CRISPR-Cas base editor that was purpose-built to bind and alter DNA was in fact able to bind and modify RNA. The same research group was able to build target discrimination back into re-engineered base editors, which performed with orders of magnitude greater fidelity compared to the first-generation designs.[28] It is likely that with engineering improvements and testing, clinical use of base editors will take off. For gene editing therapeutics, a safety range will have to be established, as with any small molecule drug or biologic. The potential for harmful genome modifications necessitates early, preclinical assessment with genotoxicity assays that cover the unique biological activities of these systems.

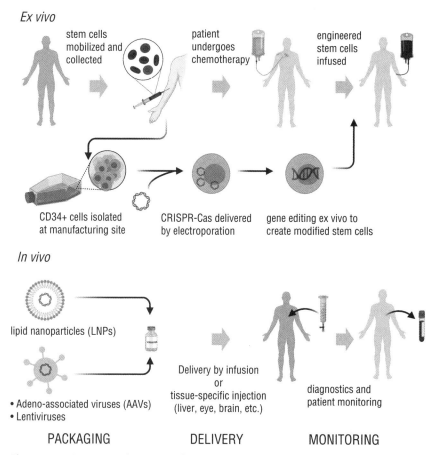

Ex vivo

stem cells mobilized and collected

patient undergoes chemotherapy

engineered stem cells infused

CD34+ cells isolated at manufacturing site

CRISPR-Cas delivered by electroporation

gene editing ex vivo to create modified stem cells

In vivo

lipid nanoparticles (LNPs)

• Adeno-associated viruses (AAVs)
• Lentiviruses

Delivery by infusion or tissue-specific injection (liver, eye, brain, etc.)

diagnostics and patient monitoring

PACKAGING   DELIVERY   MONITORING

Figure 4.6: Overview of strategies for delivery of CRISPR-Cas engineered therapies to patients

Two general approaches are currently used in investigational studies to introduce gene editing-based therapies into patients.

Top panel: Illustration of an *ex vivo* gene therapy process used primarily with production of gene-edited CAR-T cells or hematopoietic stem and progenitor cells. Delivery of CRISPR-Cas9 components is done by electroporation. (Not pictured: delivery of genes without editing is done in gene therapy trials using integrating lentiviral vectors.) The complex steps to generate edited stem cells are performed at a dedicated manufacturing facility.

Bottom panel: The delivery of CRISPR-Cas components *in vivo* is accomplished by encapsulating the DNA or RNA into lipid nanoparticles or packaged in viruses. The resulting drug formulation is administered by different routes, as indicated in the bottom panel.

Genome editing platforms will also have to grapple with the current lack of understanding around dependencies related to cellular DNA repair pathways, cell state, and cell type for a given edit. Desired genomic changes may be hampered by a cell's state (quiescent versus actively dividing) and particular cell type. For example, CRISPR-Cas approaches that depend on HDR to repair or edit sequences will need to optimize the chemistry to achieve optimal therapeutic outcomes and avoid collateral damage.

Gene editing technology is already setting a new baseline as to what is technically achievable compared to what can be done with conventional gene therapy, with an armamentarium of strategies using the newly engineered CRISPR-Cas proteins. Attacking disease with gene editing strategies can proceed in a straightforward manner with correction of point mutations in single genes, disruption of genes, or genetic loci with targeted insertions or gene knockout, or by creating novel gene modifications. Researchers have also built transcriptional control systems with transcriptional repressors and activators tethered to Cas proteins. Inducible systems for temporal control of a subset of these accessory proteins (and Cas endonuclease) have been developed, with light-activatable or small molecule-based switches.

A second layer of editing approaches aims to correct the disease phenotype via another gene whose function might substitute for the errant disease allele by complementation or even antagonism. A third group of strategies takes advantage of gene editing technology to enhance cell therapies (principally T cells and hematopoietic stem cells) that had previously employed other gene transfer approaches to engineer, for example, chimeric antigen receptors.

The daunting challenges that loomed ahead for the clinical application of CRISPR-Cas gene editing did not deter a new class of entrepreneurs to press forward. The leaders in the field, principally Jennifer Doudna, Emmanuel Charpentier, and George Church, along with Feng Zhang, established startups and began laying the groundwork. These startups, together with industry collaborators, moved rapidly toward vetting potential therapeutic scenarios for use of the technology. The Cas9 nucleases—which at the time were better understood and could be used as a gene disruption tool—were a safer first bet for gene editing in humans. The critical first decisions were made during 2014–2016 for these companies. And nearly all converged on a gene editing approach to produce a one-shot cure for hemoglobinopathies, namely, sickle cell disease (SCD) and β-thalassemia. Both diseases are caused by mutations in the gene encoding the beta (β) subunit of hemoglobin. Inheriting one mutant gene copy will produce the sickle cell trait, but individuals are typically healthy; two mutant copies cause severe disease starting sometime after birth, with lifelong health problems that include episodic pain due to vascular obstruction (a consequence of erythrocyte deformation—the sickle cell phenotype), anemia, hypertension, stroke, organ failure, and premature death.

CRISPR Therapeutics signed a research agreement with Vertex Pharmaceuticals in 2015 and viewed SCD as a promising area to trial therapeutics that relied on the ability of CRISPR-Cas9 to disrupt a carefully selected genomic target in hematopoietic stem and progenitor cells (HSPCs). What was the best attack strategy? Gene therapy clinical trials for SCD were already underway from the biotech company Bluebird Bio, which focused on gene replacement, attempting to overcome the disease by introduction of a new, functional copy of an adult β-globin gene that could counteract sickle hemoglobin. With CRISPR-Cas9, the guide RNA and Cas nuclease could be delivered to stem cells to make a targeted gene edit, in which the cellular homology-directed repair (HDR) pathway would correct the mutation in the defective β-globin sequence. However, this approach would not work in HPSCs where the HDR pathway is minimally active. The best option then was to leverage CRISPR-Cas9 to create a double-strand break at a genomic site and utilize the NHEJ pathway active in HSPCs to corrupt the targeted sequence by indel formation.

An alternative therapeutic strategy was to focus on increasing fetal hemoglobin levels in adult patients with SCD. The ability of the fetal form of hemoglobin to reduce the pathologic consequences of sickle hemoglobin had long been known,[29] and recent genomic studies offered tantalizing clues on how to increase its production in adults. During normal human development, globin genes are turned on and off sequentially, first to produce fetal hemoglobin (the protein is composed of two alpha (α) and two gamma (γ) subunits from the respective globin genes) followed by adult hemoglobin (2 α subunits and 2 β subunits). Normally the globin genes "switch" after birth, where γ-globin expression is turned off and β-globin is turned on. There are rare individuals that co-inherit pathogenic β-globin mutations causing SCD or β-thalassemia along with genetic variants that cause a condition known as hereditary persistence of fetal hemoglobin. In these people, fetal hemoglobin levels remain high into adulthood, and they experience very mild or no disease. Thus, they had inherited a way to disable the molecular control switch that was shutting down expression of the γ-globin gene. The "off" switch was discovered to be a transcriptional repressor known as

BCL11A. The protected individuals had variants in a small enhancer region that disabled the expression of this gene. The final piece was in place, giving CRISPR Therapeutics and Vertex a compelling way to use their gene editing technology. CRISPR-Cas9 would be programmed to disrupt the enhancer in the BCL11A gene, killing the "off" switch and hopefully bring fetal hemoglobin to therapeutic levels in a patient's stem cells. A similar strategy was taken by Sangamo Therapeutics using ZFNs and Intellia Therapeutics with CRISPR-Cas9. Editas Medicine went down a slightly different path to blocking BCL11A by targeting promoter sequences within the β-globin locus with CRISPR-Cas12a.

Therapeutic genome editing as a new approach for gene therapy is now on the cusp of its first drug approval. A summary of clinical trials in which gene editing was employed or currently ongoing is shown in Table 4.2. In the clinic, use of Sangamo Therapeutic's zinc finger platform preceded the emergence of CRISPR-Cas platform deployments to edit targets either *in vivo* or *ex vivo*. Some of the early results from Sangamo were disappointing from a medical standpoint, but the technology cleared critical safety evaluations. Trials investigating use of the technology in mucopolysaccharidosis types I and II (MPS I and MPS II, Table 4.2) came up short in benefiting patients, possibly due to under dosing, and other programs were dropped for various reasons. Beyond the gene editing startups, other biotech companies specializing in cell therapies have waded in, especially for oncology products, where CAR-T cells are being engineered with programmable nucleases to enhance their immunotherapeutic actions. Readouts from these oncology trials are only emerging and have yet to reveal a major therapeutic breakthrough. That hope appears to be coming from treatments for blood disorders.

In December 2020, a raft of good news was released to the public regarding the potential for CRISPR-Cas9 gene editing to be safe and effective for curing human disease. Data was presented on 10 patients at the American Society of Hematology annual meeting, and in parallel, the first scientific report on two patients treated with CTX001, the therapeutic from CRISPR Therapeutics and Vertex, was published in the *New England Journal of Medicine*.[30] From nearly everyone's vantage point in industry, it was clear just by looking at the trial results that the CRISPR technology had worked remarkably well. Preclinical assessments of CTX001's target specificity detected zero off-target edits and found the expected on-target editing activity at the BCL11A enhancer site. In the two patients, analysis of transplanted stem cells revealed that the CRISPR-Cas9 system was editing the enhancer site at a frequency of 80 percent. The question remained whether this level of genome modification was sufficient to release the "off" switch and turn on expression of the γ-globin gene to produce sufficient quantities of fetal hemoglobin to be therapeutic.

Table 4.2: Clinical Trials Using Gene Editing with Programmable Nucleases

DISEASE APPLICATION	ADMINISTRATION EX VIVO CELL TYPE OR IN VIVO	EDITING PLATFORM	GENE TARGET(S)	TARGETING RATIONALE	TRIAL SPONSOR(S)
Blood disorders					
Hemophilia B	Hepatocytes	ZFN	Factor IX	Gene replacement	Sangamo Therapeutics
β-thalassemia	CD34+ I-SPCs	ZFN	BCL11A	Gene regulation to increase fetal hemoglobin expression	Sangamo Therapeutics
		CRISPR			CRISPR Therapeutics and Vertex
Sickle cell disease	CD34+ HSPCs	ZFN			Sangamo and Sanofi
		CRISPR			CRISPR Therapeutics and Vertex
			HBG1 and HBG2 promoter regions		Intellia and Novartis
					Editas Medicine
Metabolism					
MPS I	Hepatocytes	ZFN	IDUA	Gene replacement	Sangamo Therapeutics
MPS II			IDS	Gene replacement	Sangamo Therapeutics
ATTRv-PN	in vivo intravenous administration	CRISPR	TTR	Gene disruption	Intellia Therapeutics
Oncology					
B-ALL	CAR T cells	TALEN	CD52, TRAC	Gene disruption TCR engineering	Cellectis, Servier, Allogene Therapeutics

Continues

Table 4.2 (continued)

DISEASE APPLICATION	ADMINISTRATION EX VIVO CELL TYPE OR IN VIVO	EDITING PLATFORM	GENE TARGET(S)	TARGETING RATIONALE	TRIAL SPONSOR(S)
Relapse/refractory Non-Hodgkin lymphoma	CAR T cells	CRISPR	PD-1, CD19 TRAC	Gene disruption and TCR engineering	Caribou Biosciences
			β2M, CD19 TRAC	engineering	CRISPR Therapeutics
Renal Cell carcinoma			CD70		CRISPR Therapeutics
Multiple Myeloma			BCMA		CRISPR Therapeutics
AML			CD123, TRAC		Cellectis
Multiple myeloma			PD-1, TCRendo		Tmunity
Leukemia and Lymphoma			CD7, CD28		Baylor College of Medicine
Gastrointestinal cancer	T cells	CRISPR	CISH		Intima Bioscience
Glioma	CD8+ T cells	ZFN	IL13Rα2		City of Hope Medical Center
Neurology					
LCA10	in vivo subretinal injection	CRISPR	CEP290	Gene restoration by splicing repair	Editas Medicine and Allergan
Infectious disease					
HIV-1	CD4+ T cells	ZFN	CCR5	Gene disruption	Sangamo Therapeutics

Clinical trial data includes those studying a US FDA-regulated product and listed on `Clinicaltrials.gov` as of 12/15/2020. HSPCs: hematopoietic stem and progenitor cells; iPSCs: induced pluripotent stem cells; CAR T: chimeric antigen receptor T cells; ATTR-vPN: hereditary transthyretin amyloidosis with polyneuropathy; MPS I: mucopolysaccharidosis type I; MPS II: mucopolysaccharidosis type II; TRAC: T cell receptor alpha locus; TCR: T cell receptor.

The clinical data from the first patients were nothing short of spectacular. Within a few months after infusion, both individuals were producing substantial amounts of fetal hemoglobin. There was durable engraftment of the stem cells that resulted in sustained fetal hemoglobin production out to 18 months. Prior to the study, both had experienced numerous vaso-occlusive crises, which were eliminated after a few months of treatment with CTX001. These two patients have also depended on blood transfusions over their lifetimes, and the need for transfusions abruptly stopped as well. A similar drastic change in the course of disease was reported for the other study participants (two more with SCD, six with transfusion-dependent β-thalassemia). The landmark clinical application of this bacterial defense system-based biotechnology, which was put into practice only six years prior, had already changed lives and medical history.

The biotechnology miracle drug that CTX001 might one day become affords a cure for only a tiny fraction of the more than 300,000 people globally that are born with SCD. The expensive stem cell manufacturing process and hospital care that are necessary to deliver the cure will push the price of this new medication into the million-dollar range per patient. In sub-Saharan Africa, which accounts for three out of every four of those born with SCD, 50 to 90 percent of the afflicted children will die before their fifth birthday. Even though SCD is the most common of the rare, Mendelian disorders in humans, an exquisitely designed biotechnology cure may have little impact on the disease in Africa or elsewhere. The sad reality is that the future impact of gene therapy on worldwide health is likely to be undermined by financial constraints. As with some of the more potent cancer immunotherapies, gene editing–based therapeutics may remain within the confines of affluent countries or a privilege for those with access to extraordinary wealth.

Biotechnology to the Rescue: Vaccine Development Platforms Based on Messenger RNA

What if the biotechnology industry could build an inexpensive medicine that could potentially provide life-saving benefit to billions of people during a global crisis? In one of the most stunning achievements by the scientific community of this century, the rapid development of mRNA-based vaccines for SARS-Cov-2 provided the answer. In stark contrast to the rapid innovation around CRISPR-based therapeutics, the development of a drug and vaccine platform around mRNA was decades in the making. The swift rise of CRISPR gene editing platforms followed a singular insight, whereas the potential for mRNA therapies gained traction only after key innovations were in place. To arrive at a mRNA vaccine, three major innovations had to converge: defining the chemical composition of a stable and less immunoreactive mRNA molecule, developing an efficient delivery system, and engineering of a viral protein that would generate the most potent immune response in human cells.

The recombinant DNA revolution largely left behind RNA as a source of genetic information for building drugs and vaccines. The central dogma places mRNA as the transient intermediate in information flowing from the genome to the production of proteins. RNA, especially mRNA, is a labile macromolecule, with cellular processes built for mRNA decay that rapidly degrade the signal after its transport from the nucleus to cytoplasmic ribosomes for protein synthesis. The half-life of a typical mRNA is thirty minutes to an hour, enough to make a few hundred proteins. Degradation of mRNA is by design—turnover of mRNA is a necessary control point in the regulation of gene expression. A second aspect of the molecule's instability is found in aqueous environments. RNA is prone to base-catalyzed hydrolysis, making its packaging and delivery a problem for drug formulation. DNA, the far more stable nucleic acid, is protected as it lacks the hydroxyl group found in RNA that can attack the sugar-phosphate backbone. As a result of DNA's durability, it has been sequenced from fossilized remains of plants, animals, and microorganisms in which intact RNA has never been found.

A biotechnology platform based on unstable mRNA would also have to navigate around protections set up against foreign RNAs by the immune system. To have any chance at success, the labile molecules would not only need to enter cells intact, but the artificial mRNA would also then have to escape detection intracellularly. Humans have potent innate immunity built for protection against nucleic acids found in bacteria and viruses. Antiviral responses are triggered by double-stranded RNA, which induces interferon-mediated inflammatory reactions. A single-stranded mRNA that mimicked a naturally occurring molecule might escape notice. In the late-1990s, Drew Weissman and Katalin Karikó at the University of Pennsylvania discovered, however, that there was a strong immune response against single-stranded RNA when Weissman attempted to make an HIV vaccine component from RNA synthesized in Karikó's lab. Karikó was disappointed. She had long held out dreams to use naked mRNA as therapy and was perplexed that cells were reacting against mRNA. What was the immune system detecting?

Over the next several years, the duo had begun experiments to answer that question. Surveillance and detection of foreign DNA and RNA were known to be governed by proteins known as *Toll-like receptors (TLRs)*, a highly conserved group that provide pattern recognition capabilities for pathogen-derived molecular sequences. For DNA, the structural basis for recognition of pathogen-derived sources was discovered to be a small signature—an unmethylated CpG motif (a cytosine followed by guanine with a lower case p denoting phosphate linkage). Cytosines in mammalian DNA that are found in CpG contexts are mostly methylated, whereas they are unmodified in microbial genomes. The mammalian TLR9 receptor detects this DNA motif. Karikó and Weissman had begun looking at a range of possibilities, exploring the types of RNAs that could provoke an immune response. Several RNA classes contain modified *nucleosides*, with tRNAs containing high levels of pseudouridine, an isomer of uridine that was known to stabilize the cloverleaf-like 3D structures of these molecules. In their early experiments looking for clues, tRNAs caused almost no inflammatory response or activation of dendritic cells.

In 2002, Karikó had set up a sensitive assay in monocytes to test different RNA structures for their ability to induce production and secretion of IL-12, the proinflammatory cytokine. One of the most potent inducers was synthetic RNA, which contained long stretches of uridine, polyU. The same experiment with polyA had no effect. Karikó and Weissman realized then that modified nucleosides in mRNA at positions where there were uridines might be a way to lessen or evade the immune response. In a landmark paper published in 2005, they demonstrated that RNA molecules that were synthesized with a range of modified nucleosides or pseudouridine could indeed circumvent the immune trigger.[31] This insight opened the door to refining an mRNA platform based around nucleoside engineering. Two biotechnology companies were founded around the idea, BioNTech, in 2008, and Moderna, in 2011. The early seeds of a revolutionary vaccine technology were planted.

A critical element in vaccine development programs is the choice of the *immunogen*, regardless of whether it is presented to the human immune system as an inactivated or live-attenuated virus, a viral protein subunit, DNA or RNA, or from genes in virus-vectored vaccines. As the HIV epidemic was raging across the globe in the 1990s, Anthony Fauci, head of the National Institute of Allergy and Infectious Disease, was given authority to establish the Vaccine Research Center to drive innovation in vaccine technology. The development of vaccines that were effective against respiratory viruses took on new urgency as the industry struggled to develop effective vaccines against new influenza strains, respiratory syncytial virus (RSV), and the new coronaviruses—SARS-CoV in 2003 followed by MERS-CoV in 2012. One of Fauci's deputies, Barney Graham, had a research group working on the structural basis of high-performing, neutralizing antibodies and the use of structure-based design to engineer antigens for better immunogenicity.

One of the group's important insights came in 2013 through the visualization of antibody-antigen complexes that revealed how the potency of an RSV antibody was related to an antigenic site only found on a prefusion form of the virus protein.[32] RSV enters cells by using its fusion protein to merge with cellular membranes. The fusion process results in a changed conformation of the virus protein. Any immunization strategy depending on the structure of the post-fusion antigen may not elicit an immune response if the epitope is hidden or has been lost during the conformational state change. With this knowledge, Jason McLellan, Graham, and his colleagues at the Vaccine Research Center engineered a series of fusion protein variants as immunogens for RSV. The results in animals confirmed the potency of the novel vaccine candidate. For the first time, a structure-based engineering approach had been applied successfully to vaccine development.[33]

The focus then turned to the coronaviruses, and another success was reported in 2017 when Graham and colleagues modified the spike glycoprotein of MERS with a "2P design"— placing the amino acid proline at two positions to again stabilize a prefusion conformation. Moderna, the well-financed biotech operating in stealth mode in Cambridge, had been working on mRNA-based vaccines for Zika and influenza. In 2019, the biotech had made considerable progress on its modified nucleoside technology, signed several pharma deals, and then entered into a partnership with the NIH vaccine group to work on Nipah virus. By this time, Moderna had also developed a lipid nanoparticle (LNP) delivery system that was key to getting the mRNA into cells and kept safe behind a lipid barrier.[34,35] In Germany, BioNTech was not far behind with a similar platform.

The outbreak of the SARS-CoV-2 virus in late 2019 was about to put the new mRNA vaccine platforms to the test on the world's stage. As soon as the viral genomic sequence was published on January 10, 2020, researchers had locked in on the spike (S) protein sequence as the major antigen target for vaccine development. Three days later, Moderna scientists had formulated an mRNA molecule on paper that would incorporate the MERS coronavirus 2P design (labeled S-2P in SARS-CoV-2 vaccine projects). Within 45 days, Moderna had initiated clinical trials, placing vaccine development on an incredibly rapid trajectory for human testing.

BioNTech had partnered with Pfizer and raced forward with the same strategy, using S-2P. The two biotechnology companies announced astounding phase 3 clinical trial results from the mRNA-based vaccines, with efficacy for each approaching 95 percent. The mRNA platforms were a stunning success and stood out among the 220 vaccines for SARS-CoV-2 that were under development worldwide. It was evident that it had all come together. The prefusion stabilization of the 2P design produced potent neutralizing antibodies as the modified, LNP-encased mRNAs delivered a stably translated message in cells. No less important, there was essentially no evidence in clinical trials, or from the massive vaccine rollout, of adverse immune reactions against the synthetic mRNA molecule. Biotechnology, with the backing of decades of academic and government-based research, had produced a triumph.

The utility of mRNA platforms outside of vaccines will begin to be tested soon, as a few clinical trials are underway. Moderna has a slew of therapeutics under development, and its first partnership with AstraZeneca was reported to have 40 drug candidates in process. Alexion's relationship with Moderna focuses on rare diseases. BioNTech, Merck, Moderna, and others have pushed hundreds of millions of dollars and a number of cancer vaccines into their preclinical development pipelines.

Conceptually, having cells of the body manufacture proteins from an introduced mRNA offers some advantages over recombinant protein production or traditional gene therapy. For an mRNA therapeutic, the bioreactor is the normal physiological site and production levels are controlled by cellular processes, features that favor mRNA over recombinant proteins. Gene therapies require integration into the genome, which comes with unpredictable hazards. Vaccines were ideal for initial tests of the mRNA platform, since a one or two dose injection is all that is required. The drawback for protein and mRNA-based therapeutics is the continual need for infusions, potentially over a lifetime. It is unknown whether repeated exposure to the LNPs or mRNAs will trigger unwanted immune reactions or create other toxicities. Moderna and AstraZeneca are building a heart attack drug that requires epicardial injection to deliver VEGF-A, a blood vessel growth factor. Examples of other mRNA drugs are those for cystic fibrosis in a formulation that would be inhaled (Translate Bio), and mRNAs encoding proteins for correcting proprionic acidemia (Moderna) and ornithine transcarbamylase deficiency (Arcturus) by intravenous injection.

The biotechnology innovation engine has created other molecular tools and strategies to correct genetic diseases. A return to efforts around therapies for Huntington's disease serves as an illustration. Since Huntington's is an autosomal dominant disease and the result of a single copy of an aberrant protein wreaking havoc, a commonsense strategy is simply to block its expression. CRISPR-Cas9 (Intellia Therapeutics) is being used to disrupt the Huntington gene locus and a Cas9-based RNA editing (LocanaBio) strategy employed to destroy transcripts containing toxic repeats. RNA interference strategies are being tested against the mutant transcript (UniQure's AMT-130 miRNA, Voyager's VY-HTT01 miRNA) and to control repeat expansions by knocking down a component in the DNA damage response pathway (Triplet Therapeutics).

Second-generation biotechnology tools have opened up a vast expanse of opportunities to treat disease with unparalleled precision. The exponential advance in precision is the consequence of engineering approaches toward therapeutics. Random screening and selection are being replaced by rational design and methods employing directed evolution. This is manifested in the creation of mRNA vaccines, stem and T cell engineering for immunotherapies and blood disorders, and editing of human, animal, plant, and microbial genomes for producing cures and novel therapeutics. Marvelously engineered chimeric proteins with fine-tuned properties are being created off of the Cas9 and Cas12a nucleases for high-fidelity RNA and DNA editing. The use of machine learning is aiding the selection of the best-performing base editors, and computational predictions are driving precision vaccinology. All of drug development marches forward with the innovation ecosystem of biotechnology. But progress has always been gated by the level of understanding of fundamental biological processes in health and disease. The hope is that iterative cycles of technology innovation, application of computational power and artificial intelligence tools, and greater biological knowledge will bolster continued innovation and the evaluation of new biotechnology-based therapeutics.

Recommended Reading

Doudna, J. A. and Sternberg, S. A. *A Crack in Creation: Gene Editing and the Unthinkable Power to Control Evolution*. Houghton Mifflin Harcourt: Boston, 2017.

International Commission on the Clinical Use of Human Germline Genome Editing, *Heritable Human Germline Editing*, The National Academies Press, 2020.

Notes

1. Stolberg SG. The Biotech Death of Jesse Gelsinger. The New York Times. 1999 Nov 28.

2. Pollack A. F.D.A. Halts 27 Gene Therapy Trials After Illness. The New York Times. 2003 Jan 15.

3. Hacein-Bey-Abina S, Von Kalle C, Schmidt M, McCormack MP, Wulffraat N, Leboulch P, et al. LMO2-associated clonal T cell proliferation in two patients after gene therapy for SCID-X1. Science. 2003 Oct 17;302(5644):415–9.

4. Office of the Commissioner. FDA approves novel gene therapy to treat patients with a rare form of inherited vision loss FDA. 2017. Available from: https://www.fda.gov/news-events/press-announcements/fda-approves-novel-gene-therapy-treat-patients-rare-form-inherited-vision-loss.

5. Jinek M, Chylinski K, Fonfara I, Hauer M, Doudna JA, Charpentier E. A programmable dual RNA-guided DNA endonuclease in adaptive bacterial immunity. Science. 2012 Aug 17;337(6096):816–21.

6. Doudna JA, Charpentier E. Genome editing. The new frontier of genome engineering with CRISPR-Cas9. Science. 2014 Nov 28;346(6213):1258096.

7. Donohoue PD, Barrangou R, May AP. Advances in Industrial Biotechnology Using CRISPR-Cas Systems. Trends Biotechnol. 2018 Feb;36(2):134–46.

8. Hoagland MB, Stephenson ML, Scott JF, Hecht LI, Zamecnik PC. A soluble ribonucleic acid intermediate in protein synthesis. J Biol Chem. 1958 Mar;231(1):241–57.

9. Brenner S, Jacob F, Meselson M. An Unstable Intermediate Carrying Information from Genes to Ribosomes for Protein Synthesis. Nature. 1961 May;190(4776):576–81.

10. Gros F, Hiatt H, Gilbert W, Kurland CG, Risebrough RW, Watson JD. Unstable Ribonucleic Acid Revealed by Pulse Labelling of Escherichia Coli. Nature. 1961 May;190(4776):581–5.

11. Crick FHC, Barnett L, Brenner S, Watts-Tobin RJ. General Nature of the Genetic Code for Proteins. Nature. 1961 Dec;192 (4809):1227–32.

12. Berg P, Baltimore D, Brenner S, Roblin RO, Singer MF. Asilomar conference on recombinant DNA molecules. Science. 1975 Jun 6;188(4192):991–4.

13. Jackson DA, Symons RH, Berg P. Biochemical Method for Inserting New Genetic Information into DNA of Simian Virus 40: Circular SV40 DNA Molecules Containing Lambda Phage Genes and the Galactose Operon of Escherichia coli. Proc Natl Acad Sci. 1972 Oct 1;69(10):2904–9.

14. Cohen SN, Chang ACY, Hsu L. Nonchromosomal Antibiotic Resistance in Bacteria: Genetic Transformation of Escherichia coli by R-Factor DNA. Proc Natl Acad Sci. 1972 Aug 1;69(8):2110–4.

15. Cohen SN, Chang ACY, Boyer HW, Helling RB. Construction of Biologically Functional Bacterial Plasmids in vitro. Proc Natl Acad Sci. 1973 Nov 1;70(11):3240–4.

16. Singer M, Soll D. Guidelines for DNA Hybrid Molecules. Science. 1973 Sep 21;181(4105):1114–1114.

17. Morrow JF, Goodman HM, Helling RB. Replication and Transcription of Eukaryotic DNA in Escherichia coli. Proc Nat Acad Sci USA. 1974;5.

18. The Huntington's Disease Collaborative Research Group. A novel gene containing a trinucleotide repeat that is expanded and unstable on Huntington's disease chromosomes. Cell. 1993 Mar 26;72(6):971–83.

19. Campbell PJ, Getz G, Korbel JO, Stuart JM, Jennings JL, Stein LD, et al. Pan-cancer analysis of whole genomes. Nature. 2020 Feb;578(7793):82–93.

20. Stenton SL, Prokisch H. Genetics of mitochondrial diseases: Identifying mutations to help diagnosis. EBioMedicine. 2020 Jun 1;56.

21. Jansen R, Embden JDA van, Gaastra W, Schouls LM. Identification of genes that are associated with DNA repeats in prokaryotes. Mol Microbiol. 2002;43(6):1565–75.

22. Bolotin A, Quinquis B, Sorokin A, Ehrlich SD. Clustered regularly interspaced short palindrome repeats (CRISPRs) have spacers of extrachromosomal origin. Microbiol Read Engl. 2005 Aug;151(Pt 8):2551–61.

23. Mojica FJM, Díez-Villaseñor C, García-Martínez J, Soria E. Intervening sequences of regularly spaced prokaryotic repeats derive from foreign genetic elements. J Mol Evol. 2005 Feb;60(2):174–82.

24. Pourcel C, Salvignol G, Vergnaud G. CRISPR elements in Yersinia pestis acquire new repeats by preferential uptake of bacteriophage DNA, and provide additional tools for evolutionary studies. Microbiol Read Engl. 2005 Mar;151(Pt 3):653–63.

25. Barrangou R, Fremaux C, Deveau H, Richards M, Boyaval P, Moineau S, et al. CRISPR Provides Acquired Resistance Against Viruses in Prokaryotes. Science. 2007 Mar 23;315(5819):1709–12.

26. Cong L, Ran FA, Cox D, Lin S, Barretto R, Habib N, et al. Multiplex Genome Engineering Using CRISPR/Cas Systems. Science. 2013 Feb 15;339(6121):819–23.

27. Ledford H. CRISPR gene editing in human embryos wreaks chromosomal mayhem. Nature. 2020 Jun 25;583(7814):17–8.

28. Grünewald J, Zhou R, Garcia SP, Iyer S, Lareau CA, Aryee MJ, et al. Transcriptome-wide off-target RNA editing induced by CRISPR-guided DNA base editors. Nature. 2019 May;569(7756):433–7.

29. Watson, Janet J. A study of sickling of young erthrocytes in sickle cell anemia. Blood. 1948 Apr 1;3(4):465–9.

30. Frangoul H, Altshuler D, Cappellini MD, Chen Y-S, Domm J, Eustace BK, et al. CRISPR-Cas9 Gene Editing for Sickle Cell Disease and β-Thalassemia. N Engl J Med. 2020 Dec 5.

31. Karikó K, Buckstein M, Ni H, Weissman D. Suppression of RNA Recognition by Toll-like Receptors: The Impact of Nucleoside Modification and the Evolutionary Origin of RNA. Immunity. 2005 Aug 1;23(2):165–75.

32. McLellan JS, Chen M, Leung S, Graepel KW, Du X, Yang Y, et al. Structure of RSV Fusion Glycoprotein Trimer Bound to a Prefusion-Specific Neutralizing Antibody. Science. 2013 May 31;340(6136): 1113–7.

33. McLellan JS, Chen M, Joyce MG, Sastry M, Stewart-Jones GBE, Yang Y, et al. Structure-Based Design of a Fusion Glycoprotein Vaccine for Respiratory Syncytial Virus. Science. 2013 Nov 1;342(6158):592–8.

34. Kose N, Fox JM, Sapparapu G, Bombardi R, Tennekoon RN, Silva AD de, et al. A lipid-encapsulated mRNA encoding a potently neutralizing human monoclonal antibody protects against chikungunya infection. Sci Immunol. 2019 May 17;4(35).

35. Liang F, Lindgren G, Lin A, Thompson EA, Ols S, Röhss J, et al. Efficient Targeting and Activation of Antigen-Presenting Cells In Vivo after Modified mRNA Vaccine Administration in Rhesus Macaques. Mol Ther. 2017 Dec 6;25(12):2635–47.

Healthcare and the Entrance of the Technology Titans

Among the most important and difficult challenges facing humanity in the 21st century is access to healthcare. Global public health crises such as the COVID-19 pandemic underscore the importance of medicines, innovation, access, delivery, and the economic and human cost of healthcare system failures. Regardless of whether a nation's citizens have the luxury to choose private or public options or to gain access through universal programs, the delivery of services and the maintenance of healthcare systems is enormously complex and burdensome. In the United States, relentless increases in healthcare costs leave families poorer, millions uninsured, and many patients out of reach of essential services and medications. On the provider side, primary-care physicians are forced into a one-size, 15-minute appointment-fits-all paradigm with little time to listen to patients or enter their health data into antiquated electronic health record (EHR) systems. Societal pressure and government policy changes could greatly impact many of the factors that chronically plague healthcare. Another route to change is through industry and the restructuring of incentives to help lower costs. Of course, there is also the promise of innovation—finding better ways to deliver healthcare, discovering new medicines more cheaply, or even rebuilding the entire healthcare ecosystem and its technology infrastructure.

Scientific innovation has been a decisive factor driving the growing importance and economic value of drug companies within the healthcare sector. The industry has brought a substantial number of therapies to populations that had previously suffered from common ailments such as arthritis, asthma, and angina, but healthcare expenses rose substantially with each new product introduction. Biomedical research and biotechnology breakthroughs provided the fuel to power the young industry's drug development pipelines and to produce innovative medicines in new markets. Pharmaceutical firms, on the other hand, perfected the blockbuster drug model long ago, having cemented a reliance on small molecule medicinal chemistry combined with standardized drug manufacturing processes. Although there has been a generational change toward incorporation of biotechnology-based innovation, precision medicine, and biologics as drugs, the Big Pharma business model remains intact and entrenched across the pantheon of old guard companies.

The healthcare sector is a 10 trillion-dollar global industry that includes pharmaceutical, biotechnology, and life sciences companies grouped as one subdivision and another comprising healthcare services such as hospitals, home healthcare providers, and managed care, plus medical supplies and equipment. In developed countries across Europe, the United States, and Asia, pharmaceutical spending accounts for 10–20 percent of overall healthcare expenditures. The Pharma industry as a whole has fought hard to resist efforts at price control, nearly always using the justification that research costs should be factored into the price structure, in addition to the patient's benefit.

Gleevec, the miracle cancer drug for chronic myelogenous leukemia, is a good example. At its launch in 2001, the drug was priced at $26,000 a year and rose to $120,000 annually at its patent expiration date in 2015 (which had been extended two or more years by Novartis with additional patents and an agreement with the first generic manufacturer, Sun Pharmaceuticals, to delay generic competition). Specialty drugs like Gleevec are one of the biggest drivers of increasing healthcare costs and will continue to be moneymakers for the industry well into the foreseeable future.

Prior to the start of the COVID-19 pandemic, a majority of the healthcare industry's profits were garnered by just ten companies, nine of which were pharmaceutical manufacturers. Net profit margins of pharmaceutical and biotechnology companies in the first quarter of 2019 were typical for those industries, ranging from 18.7 percent for Johnson & Johnson ($20 billion in Q1 revenue) to 83 percent for Lilly ($5 billion in revenue). By contrast, pharmacy retailing giant CVS Health recorded a net profit margin of only 2.3 percent on $61 billion in revenue in Q1 2019.

The profit margins of the biggest health insurers in the United States were also in the single digits, with UnitedHealth Group reporting 5.7 percent and Cigna at 3.6 percent in the same quarter. The pharmaceutical industry still has a firm grip on its pricing power and operational efficiencies, brought by patent protection and manufacturing at scale, respectively. Viewed from afar, the drug business has a mass production aura about its efficiency, similar to a semiconductor foundry or an automotive manufacturing plant. As detailed previously in Chapter 3, "The Long Road to New Medicines," nearly all of the top pharmaceutical companies of today were founded more than 150 years ago and remain profit machines with strong competitive barriers to entry. It is hard to see how that status might change anytime soon. So where else might healthcare be changed?

At the beginning of the 21st century, there seemed to be little in the way of disruptive forces that could step in and alter the dynamic from outside the healthcare sector and its entangled ecosystem. The immovable object needed to be confronted with an unstoppable force. Where would that come from? It is now clear in retrospect that a challenge would ascend from the tech sector. One of the megatrends of the 2000s was industrial transformation occurring as a result of the information revolution. Almost in parallel, the Internet flourished, mobile devices proliferated, and computing became cheap, virtual, and ubiquitous. Digital-only business platforms arose out of social networking, and everything pointed to information technology as a wellspring of disruption. In industries ranging from banking and finance, media and communications, transportation and energy, and especially retail, new technologies and new ways of doing business in the digital age were overturning conventional approaches. The ability of small innovators as well as large, digitally savvy corporations to reach the masses or develop services and products for healthcare providers was suddenly a possibility.

Digital Health and the New Healthcare Investment Arena

A tremendous amount of capital has flowed into new opportunities in *digital health*, as investors and entrepreneurs envisioned a tsunami of technology-driven applications for healthcare. Figure 5.1 depicts an overview of how digital health technologies are being positioned in healthcare. Money has poured in from several sources to fund thousands of investments over the past 10 years. Much of the activity is led by traditional and corporate venture capital, which together are responsible for more than 3,000 investments in thousands of companies—about 80 percent of the total activity.[1] Private equity has also taken a keen interest (215 deals, 4.9 percent) and is among the largest dealmakers. Angel groups, accelerators and incubators, hedge funds or asset managers, foundations, and health systems also participated in the growing frenzy of activity in the new subsector. The excitement has focused on market opportunities for wearable sensors, mobile device health applications, telehealth platforms, and personalized medicine tools. Although a large number of these deals went to devices, a significant portion went to analytics and business software, where tools were built for practice management on the provider side and for analysis of claims, risk metrics, and other data for payers. It turns out that growth in digital health financing far exceeded the growth of total venture capital funding (166 percent) and total number of venture capital deals (50 percent) from all other fields. The money continued to stream in from 2017 through 2020, with venture capital investments in digital health averaging between $8 to $12 billion a year.

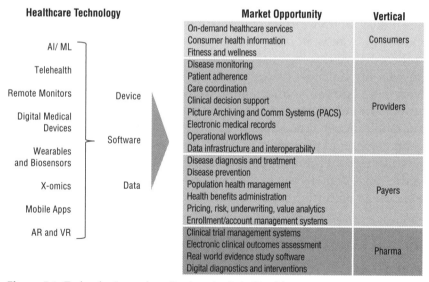

Figure 5.1: Technologies and applications in digital health

A whole range of healthcare technologies (left) are being employed either alone or in combination in the form of devices, software, and data to create digital health tools for healthcare. The market application areas (right) are segmented by market verticals with some overlap (for example, disease diagnostics and prevention cut across consumer, provider, and payer verticals).

X-Omics: Any of the molecular profiling technologies that include genomics, epigenomics, proteomics, metabolomics, and lipidomics.

The digital health arena, comprised primarily of startups, has seen some successes along with a substantial pile of failures. Among the many unicorns[2] are less well-known companies such as Tempus, which is seeking to provide AI and genomics-based treatment guidance in cancer, and Butterfly, a maker of a new ultrasound device. Flatiron Health, an entrant into the EHR space that builds software and captures data from a variety of sources around oncology, was acquired by Roche for $1.9 billion. A notable unicorn failure is Proteus Digital Health, a startup that designed both wearables and ingestible biosensors to support drug regimen adherence. The company had a valuation of more than $1.5 billion shortly before filing for bankruptcy in 2020. Of digital health ventures having raised more than $100 million, the flameouts include uBiome (microbiome testing) and Outcome Health, both suspected of dubious product claims. Early entrants into telehealth, such as HealthSpot, also ran into trouble with poor business models or adoption.

In the startup world of Silicon Valley, with its "move fast and break things" mantra and minimum viable product (MVP) model, relatively few venture-backed digital health companies have been successful. Industry experts point to failures of strategy and execution, but there are many obstacles surrounding a path to disrupting healthcare. For instance, the tech-savvy entrepreneurs have had difficulty integrating their solutions into existing healthcare technology infrastructure. Another point of failure has been the attempt to apply an outdated business model from building the second wave of Internet companies to current problems. Setting up consumer-facing health apps using a SnapChat or Uber model in today's environment is not going to work. A final point is that skepticism by doctors and health systems toward technology has been a huge factor in blocking adoption.

The renewed activity in healthcare investments has not been limited to venture-backed startups. Private equity capital investment within the medical sector in the United States has grown enormously, rising 20-fold from $5 billion in 2000 to $100 billion in 2018. The activity in 2018 was composed of 855 deals across a wide variety of healthcare services. In 2019 alone, private equity deals amounted to more than $450 billion. Private equity firms operate with a buy-to-sell business model targeting their capital resources and management expertise toward undervalued or underperforming firms in hopes of a short-term payout. This source of private capital is put toward the raw pursuit of profits, not for revolutionizing healthcare. Nevertheless, it has altered the composition of corporate entities that comprise the healthcare system with further integration and consolidation.

Across the medical landscape, new technologies and innovative business practices share the implicit promise that care will improve along many vectors. The hope is that they will address the twin goals of more affordable services and a focus on preventing and managing chronic conditions, which are responsible for 90 percent of the costs in healthcare.

DRIVERS OF HEALTHCARE TECHNOLOGY INNOVATION

The following is a summary breakdown of the key drivers for technological innovation in healthcare:

- Opportunities abound for technology to address affordability, access, patient outcomes, quality, and patient experience in health systems. Healthcare sector inefficiencies created these challenges, which in turn provide incentives for innovation, either from the outside or by incumbents, to solve any of these major problems to deliver patient-centric care.

- Payers and providers are under enormous pressure to reduce costs and improve operational efficiency. Healthcare incumbents and new entrants have a huge opportunity to leverage innovation and improve balance sheets while lowering the cost and improving the quality of healthcare.

- Precision medicine initiatives are creating opportunities for new technologies to enable this model and bridge the gap between population medicine and individualized care.

- New streams of data from pharmaceutical companies and diagnostics labs with molecular profiling technologies are increasingly a driver, especially when used in precision medicine applications.

- Healthcare technology investors and corporations are providing significant capital resources to spur innovation. Digital health has been a catalyst, and investors see big opportunities around patient engagement, AI-based analytics, cloud-based services, and new care models.

- Healthcare industry's largest players are using mergers and acquisitions as a strategy to add new technology. Payers, providers, healthcare services, and technology firms have acquired assets to strengthen data and analytics capabilities, primarily around electronic health records.

- Regulatory requirements mandated by the 21st century Cures Act in the United States create opportunities around interoperability of electronic health record (EHR) data and increase data transparency and access for consumers.

> ▪ The world's largest technology companies (based in the United States) are competing for entrance into many healthcare segments with a broad range of technological capabilities, consumer health-related products, and services to incumbents. The tech giants have formed many strategic partnerships, have completed acquisitions, and have opened new business units to exploit the vast range of opportunities in the global healthcare economy.

Assessing the Tech Titans as Disruptors in Healthcare

The global technology economy is similar in scale to healthcare and dominated by five corporations from the United States: Microsoft, Apple, Amazon, Google, and Facebook. The first four of these organizations are among the top five most valuable enterprises on the planet (Saudi Aramco is the largest). At the beginning of 2021, these tech companies combined were worth a total of $7.5 trillion based on market capitalization. The tech giants are on average 30 years old and have far eclipsed the 100-year-old dinosaurs that once ruled corporate America—the automakers GM and Ford and the fossil fuel extractors Chevron, Texaco, and ExxonMobil. The technology-centric group has come to control major chunks of the global economy, leading media (Facebook, Google), e-commerce and retail (Amazon and Apple), and software (Microsoft). In the tech economy, they are fighting for supremacy in cloud computing, which at present is led by Amazon Web Services (AWS), Google Cloud, and Microsoft Azure. Can any of these companies cross the Rubicon and play a decisive role in healthcare?

To answer the question whether the tech giants are equipped to move into healthcare, it is informative to consider how they each got to where they are today in diverse sectors of the economy. The bottom line is that over time these companies adapted in response to opportunities, evolved, and then ultimately became something else. Amazon began as an e-commerce retailer and morphed its computing infrastructure into a separate cloud computing business that is currently unrivaled as a service provider. Apple Computer Inc. became Apple Inc. in 2007, as Steve Jobs turned the company into a luxury brand offering consumer electronics beyond desktop computing. Google came onto the technology scene in the late 1990s as the Internet's 18th search engine, a far cry from having a first-mover advantage. Over time, its algorithms proved superior and enabled a new business model on a platform that eroded traditional media's relevance in advertising. Within a decade, Google, together with Facebook, held monopolistic control over Adtech and digital advertising dollars. Microsoft is the only one of the big five that remained relatively true to its roots. From MS-DOS to Windows, Microsoft has always developed and sold computer operating software, and it now reigns over the kingdom of productivity software for consumers and enterprises alike. Amazon, Apple, Alphabet (Google's parent), and Microsoft have shown considerable interest in healthcare and have placed enormous investments in health and medical applications. The industry-agnostic nature of information technology that these tech giants are built on (and control) means that nearly any sector is fair game. Can technology and the tech giants then address all of the pressing needs in healthcare? The answer is a qualified yes, as long as they play outside the boundary of practicing medicine and play the provider role.

As an example, consider the trend toward home-based care. The tech companies have support technologies for nearly all aspects of a patient-oriented care model. Each of the techs has a voice recognition technology—Amazon's Alexa, Apple's Siri, Microsoft's Cortana, and Google's cleverly named Google Assistant. With mobile and smart home devices, consumers can interact with caregivers. Communication is facilitated by patient portals and thousands of medical or health-related apps on mobile devices. Telehealth, wearables, smart clothes, edge computing, handheld devices, delivery of drugs, services—all of these have a tech component with connections to technology company platforms that can be plugged into healthcare systems. These young companies all possess boldness, global reach, and technologies that can impact healthcare in profound ways. For a few of these tech giants, their relationships with consumers might be the decisive factor in how far a technology-based corporation can go to provide products and services and lead transformational change for the healthcare sector.

Alphabet: Extending Its Tentacles Into Healthcare with Google and Other Bets

Among the tech giants, Alphabet has the longest history and most diversified strategy in approaching the healthcare sector. It is also the scariest from a privacy perspective. Google's incredible technical infrastructure, AI capabilities, and relentless pursuit of organizing the world's information would appear to make medical data and health records a core component of the company's long-term strategy in healthcare. Sundar Pichai, Alphabet

Inc. and Google's CEO, reasserted the importance of healthcare for the organization at the World Economic Forum's annual meeting in Davos in 2020.[3] Pichai believes that the company can leverage the strong privacy regulations that are already in place within healthcare to move forward with its vision. It will come down to trust of the tech giants by consumers, health organizations, and other stakeholders to enable a tech-driven transformation of the healthcare economy.

Google's first foray into healthcare came out of a patient health records project that begun quietly in 2006 and was formally launched in 2008, known as *Google Health*. The idea was to provide consumers with a portal where personal health data and medical records could be stored, searched, and shared. The service did not gain traction in the wider consumer marketplace, and the project was shut down in 2012. A similar sort of flop was repeated by Microsoft with its HealthVault offering, along with scores of other startups promising secure, web-based storage and organization of health records. The problem in part was timing—consumer appetite and readiness—not Google's service. The mainstay products of Google, including Search, Google Maps, Gmail, Google Cloud, Android, and of course Ads, were phenomenally successful. Sergey Brin and Larry Page, the cofounders of Google, had set out in 1998 to tackle a new frontier with information technology in original, creative ways, and exploratory projects like Google Health have been part of the company's DNA from the beginning. They were exuberant visionaries, and Google Health was but one of many more health projects to be piloted.

The root of some of the most prominent examples of Alphabet's belief in healthcare arose from Google X, now simply known as X, the moonshot factory established in secrecy around 2010 by the founders themselves. Headline-grabbing, audacious projects were nurtured by Brin and Page inside the laboratory, which bears little resemblance to conventional corporate research labs in Silicon Valley. The self-driving car project, Chauffeur, was the earliest project idea and led by roboticist and AI pioneer Sebastian Thrun. Out of Google X have come Google Brain, which developed deep learning capabilities from projects led by Jeff Dean, Andrew Ng, and Greg Corrado, and Verily, a life sciences company that became the primary healthcare business when Google was restructured as part of Alphabet in 2015.

Verily has established numerous collaborations with health organizations and pharmaceutical companies, developing digital tools for clinical research and clinical trials. Much of this activity centers on Project Baseline, an effort to evaluate tools for longitudinal population health studies with deals signed in 2019 with Novartis, Sanofi, Otsuka, and Pfizer.

Another partnership was set up to deploy Verily's analytics on Dexcom's G7 glucose monitoring device. Verily also combined with Google Brain researchers to help launch its AI-driven diagnostic tool for diabetic retinopathy (detailed in Chapter 6, "AI-Based Algorithms in Biology and Medicine"). Far more ambitious plans to become a significant player in its own right in healthcare are evidenced by enormous fundraising rounds, which cumulatively have totaled $2.5 billion as of early 2021. Verily's business has expanded into platforms for health insurance, chronic disease management, diabetes projects with Onduo, and its own internal research programs for radiology, immunology, and digital surgery.

Strategic investments have also played a role in furthering Alphabet's ambitions around healthcare markets. Google Ventures (GV), founded by Bill Maris in 2009, put together the biotech company and Bet subsidiary Calico. The company hired Genentech veteran Art Levinson for research into anti-aging and the development of potential cures for aging disorders. Over the past few years, GV has invested in Foundation Medicine, Genomics Medicine Ireland, Editas Medicine, Gritstone Oncology, and Flatiron Health within the healthcare sector. In addition to GV, Alphabet has its own corporate venture group and two other investment arms as Bets: Gradient Ventures, focused on AI, and CapitalG, an independent growth fund. The sprawling investment base provides an important surveillance network for the parent company, with front-row vantage points for watching new ideas develop, mutate, and materialize across health and tech startups with the latest technology and business innovations.

Acquisitions made by Alphabet paint the clearest picture of the road ahead for the tech giant in healthcare. At the end of 2014, the company announced plans to purchase DeepMind to bolster its AI technology portfolio. Alphabet funded DeepMind's expensive R&D efforts and provided it with a great deal of autonomy, but then absorbed the subsidiary's healthcare products in 2018. A second move, through the $3.2 billion acquisition of Nest in 2014 (a GV investment), brought Google into households in a unique way. And business activities of Nest, namely, its acquisition of Senosis Health in 2018, revealed that Alphabet and its subsidiary were positioning the Nest smart appliance to be part of an in-home digital health solution. Senosis Health designs mobile health monitoring apps. Integration of its software products with Nest appliances and Google voice assistance technology could serve as an at-home health supporting device for the "aging in place" market. More recently, the finalization of Google's purchase of Fitbit for $2.1 billion gives Alphabet a competitive wearable for taking on the Apple Watch and staying ahead of Amazon's Halo in fitness and wellness categories. Another stealthy maneuver was the acquisition of Apigee for its application programming interface (API) management platform in 2016. While thought of as a move to differentiate the Google Cloud platform from Amazon's AWS for corporate customers, Apigee's technologies could be a game-changer for interoperability in healthcare.

Google Health was revived in 2018 with new leadership and has again placed emphasis on the central role of healthcare data in its plans. The group's well-publicized patient data privacy breach (some Google employees had access to personal health information) from a collaboration with Ascension Health provided a window to view its strategy. Google Health's product development and commercialization efforts include a medical data platform with ambitions not only to improve search across records in EHR systems such as Epic's, but also to provide the health data infrastructure layer. Another agreement with UCSF was designed to allow Google to get free access to data to test analytics across certain types of health records. They are also working on a clinical assistant app called Streams from DeepMind Health. The Google Health research team will benefit broadly from the Fitbit acquisition. Using Fitbit's devices, the team's clinical research can expand into cardiovascular, sleep, and respiratory health.

Google Cloud now offers an array of tools to organize health data and medical research information, address medical problems with AI technology, and tackle digital health data in large EHR systems. The engineers at Google Cloud have created an AI-based tool for scanning clinical notes called Healthcare Natural Language API. A companion product, AutoML Entity Extraction, gives clinicians machine learning tools to search the records for relevant information and derive medical insights. These tools are a magnet to draw providers into Google Cloud and Google's Cloud Healthcare API service, although users can bypass the service. This is also a direct competitor with Amazon's HealthLake.

Google and the subsidiaries Verily, Calico, and DeepMind have planted a flag into the ground to claim a role in healthcare. The various elements of Alphabet have brought together clinical hardware, clinical decision support systems, and clinical research; health system infrastructure software, analytic tools, and databases; and application software together with new medical devices and wearables. With Fitbit, the company has gained entry into more than 40 Medicare Advantage plans across 27 states and large insurance firms like UnitedHealth. Any casual observer would notice the depth and variety of projects around healthcare that Alphabet has made. To date, however, the company's main source of revenue is almost entirely in advertising. In the first quarter of 2020, Google reported $40.9 billion in revenues from advertising; all of the other Bets combined only provided $135 million. It is unclear what would trigger the parent company to establish a business focus on health specifically. The business model currently seems an awkward fit for engaging with consumers around health or operating services for providers. Only time will tell.

Apple Inc: Consumer Technology Meets Healthcare

If you zoomed out into the future, and you look back and you ask the question, "What was Apple's greatest contribution to mankind?"—it will be about health.

—Tim Cook, in an interview with Jim Cramer on CNBC television January 2019

The technology and economic forces that propelled Apple to become the world's first trillion-dollar company stand in stark contrast to Google's platform for advertising built on top of Internet search algorithms. Steve Jobs turned Apple Computer from a hardware company into Apple Inc., a consumer luxury brand delivering dazzling products and a cult-like user experience to its legion of followers. Apple has built its reputation and loyalty around user's affinity for and trust in its products and a corporate orientation for protecting user privacy. Unlike Google, Facebook, and Amazon, Apple does not rely on sending user data and content into the cloud—ultimately for monetization. In comparing data privacy practices and business models, the two tech giants—Apple and Google—come out on opposite sides of the spectrum. Inside the corridors of Apple's spaceship-style headquarters at 1 Infinite Loop in Cupertino, California, are hardware and software engineers huddled together with corporate executives that are obsessed with product engineering and design, not in gathering and analyzing information as a business. Thus, Apple has approached healthcare through a much different lens, seeing opportunities primarily from the point of view of the consumer.

The previous quote from Tim Cook, Apple's current CEO, may seem surprising on its surface. If healthcare is an aspirational goal for Apple—its lasting contribution to humanity—why spend so much effort on making the best pocket-sized camera instead of a medical device or a cancer detection algorithm? Apple to date has not gone down the road of FDA-regulated medical devices, telehealth, or health payments. The most obvious strategic play has been to build features into its iPhone and Apple Watch products that appeal to a massive base of healthy users. Consistent with that approach, the company's merger and acquisition activities of the past two decades show a strong commitment to finding and incorporating technologies that strengthen their products. Recent acquisitions of Xnor.ai, Voysis, NextVR, and Curious AI together add new AI capabilities for the iPhone and an anticipated augmented reality/virtual reality (AR/VR) product in 2022. The quest for growth or expansion into healthcare or other sectors or verticals is not evident in any of Apple's publicly disclosed deals. The product roadmap into Apple's future outside of phones and computing devices appears to curve around wearables, AR and VR, and autonomous vehicles, with smart homes, cybersecurity, and media also in the mix.

Apple's health and wellness focus began publicly in 2014, with the inclusion of Apple's Health app on iPhones and iPod Touch devices. The timing was relatively late into the mobile and digital health arena. Or was it? Undoubtedly, Apple's Health app was developed and then situated alongside other core phone apps as a result of enthusiasm by developers, consumers, and health practitioners that had already built their own apps for placement within Apple's App Store. Consumers were already signaling to the company new terrain to conquer.

The ease of mobile application development turned out to be the superhighway for getting all sorts of medically focused ideas onto personal devices and into the economy, as well as into mainstream consciousness. The iPhone software development kit (SDK) was released on March 6, 2008, effectively timestamping the birth of mobile health. Ironically, one of the medical apps to appear in the App Store in 2009 was called Health Cloud, created by an iPhone developer to enable Google Health users to access and view their electronic medical records from an Apple product via a Google API. At a preview of the release of iPhone 3.0 in 2009, Apple's senior vice president of iPhone Software, Scott Forstall, had already anticipated that Apple could soon be providing a new class of services for medical devices. With either Bluetooth technology or USB connectivity, the new iOS could allow app developers to sync data coming from medical devices, such as blood pressure monitors or electrocardiogram (ECG) recorders. As a publicity stunt, Forstall invited a person from Johnson & Johnson's Lifescan subsidiary onstage to demonstrate how a Bluetooth-enabled blood glucose monitor could send information to a diabetes management application running on a remote iPhone. Since then, medical and health applications have seen astronomical growth on the Apple platform. After the creation of the Medical category and introduction of the App Store in 2008, the number of healthcare apps had risen from 82 to 46,608 in 2020, according to a report by Statista.[4] The consumer appetite for health monitoring, diet and exercise tracking, and medical support was certainly not left unnoticed by Apple's product teams and business strategists.

At Apple, the company's chief operating officer is Jeff Williams and the executive responsible for health teams. Williams was a key figure behind the development of the Apple Watch. As Apple realized that its marketing campaigns around the watch as a fashion accessory were not performing, he pivoted the product sharply toward fitness, pushing development of sensors and algorithms that could bolster health and wellness functions on the wearable. The gambit paid off with surging sales tied in part to apps for sleep tracking, heart rate monitoring, ECG, atrial defibrillation, and most recently blood oxygen sensing to wed the product more closely to health and fitness. There are also new SDKs for developers on the iPhone and Watch to create health-related apps: these are HealthKit, ResearchKit (medical research and clinical trials), and CareKit (for connecting patients with providers). For healthcare, this is a big win, as a world-class product developer is pushing digital health products with services attached deep into the mass market.

A second piece of Apple's healthcare strategy is allowing all of their customers to get connected to electronic health records directly through an iPhone. Apple appears to be much better at this than Google and has given the consumer a number of safe routes to access their personal health information. The Health app already has a Health Records option for connecting individuals to their health system. Apple has agreements for medical record access with athenahealth, plus Cerner and Epic, two of the largest EHR vendors. And health providers can send lab test results, treatment plans, notes on drug usage, or other data directly to a patient's device as well. With 1.5 billion active devices and a reach of more than half of the United States population, Apple is incredibly well-positioned to dominate this facet of healthcare by plugging into these complex EHR systems that control the flow and storage of medical information across the health system universe.

While Apple's consumer-facing products are a natural fit to the masses, the company is engaging with other stakeholders in the healthcare value chain. This includes initiatives with health insurance customers, such as Aetna and UnitedHealthcare, private Medicare insurers, Medicare Advantage plans, as well as pharmaceutical companies such as Eli Lilly and Johnson & Johnson. Insurers have announced plans to offer Apple Watches at discounted prices as a benefit to their customers—a win for Apple in the race to capture both insurers and their beneficiaries. Apple has also moved into providing health clinics for employees through its subsidiary AC Wellness.

Medical research efforts by Apple have accelerated as a result of the quality of the sensors and versatility of iPhones in clinical studies.

The marquee example is the Apple Heart study done with Stanford Cardiovascular Health, which recruited a whopping 419,297 participants in the space of eight months.[5] The study, done to examine atrial fibrillation, detectable by irregular pulse monitoring, used Apple's iPhone and Watch as devices, plus tools from the ResearchKit to recruit participants. Tim Cook told *Fortune* magazine that ResearchKit was made free to the research community to enable these types of enormous clinical studies. Johnson & Johnson is now conducting a clinical trial using Apple's atrial fibrillation detector. Studies have also been initiated with collaborators at Duke and Stanford to evaluate Apple products for long-term monitoring of chronic conditions.

Given Apple's product ecosystem and consumer reach, it comes as a surprise that bolder steps have not been taken to capitalize on the competitive leverage they possess for moving into other aspects of healthcare. Perhaps there is a hesitancy to seek out customer-generated health data in a population-scale manner that would be needed to make an impact on preventive medicine. The brakes might be applied due to data privacy concerns that could impact Apple's brand. It is also conceivable that the infrastructure requirements to store and analyze such vast quantities of data is not a technological strength of the company in the way that it is for Google or Amazon, or Apple may be prudently staying out of the way of regulators while healthcare policy and health systems evolve.

Out of all of the tech giants, Apple could potentially pull off something that could be truly disruptive in healthcare—to be the driving force behind a patient-centric, decentralized healthcare model. Rather than compete with centralized infrastructure, monolithic EHR systems, and a top-down approach to healthcare management, Apple could prove a testing ground for maintaining data privacy and placing the patient-physician (or care team) relationship at the center. By using Apple's AI and software technologies on wearables and IoT devices, personal data could be analyzed and interpreted locally, with results shared by use of blockchain or encryption methods with designated healthcare payers and providers. The provider in turn could request access to diagnostic results, financial and benefits information, or other data from payers and other organizations. Such a decentralized model would make providers the integration point of healthcare delivery, a powerful but more burdensome role on already overworked physicians and staff—a position that could be made far easier with well-engineered Apple products and services. Decentralization movements in other sectors, specifically aimed at restructuring social networking, news media, and banking systems, could very well spill over into healthcare where privacy concerns are paramount.

Amazon: Taking Logistics to the Next Level for Delivering Healthcare

> *As a company, one of our greatest cultural strengths is accepting the fact that if you're going to invent, you're going to disrupt. A lot of entrenched interests are not going to like it.*
>
> —Jeff Bezos, in an interview with Stephen Levy for *Wired*, 2011

Jeff Bezos left his job as a hedge fund analyst and drove across America in 1994, in search of financing for his business idea, an online bookstore he called Amazon. Bezos wasn't the first to begin an e-commerce business, but the founder's visionary approach meant that retail would never again be the same. Time and time again, Bezos has used innovative business practices rather than uber-sophisticated technology to plot takeovers of large swaths of the global economy. Amazon's superpower is in tackling gargantuan problems and deploying immense capital resources at such a scale that no other organization would conceivably want to attempt such a maneuver or have the appetite to compete against the formidable company. In traditional corporate boardrooms, the prospect of staggering operating losses and years of uncertainty are not part of the playbook in seeking a competitive edge. The process is typically framed as "what is the greatest advantage that we can build for the smallest investment and lowest risk?" For Amazon, solving a huge problem that no one else can afford to tackle was, and still is, the game plan. Providing customers with a frictionless shopping experience for obtaining goods in every conceivable product category, at low cost, and with free shipping proved costly and took two decades for the company to realize the full benefit. Amazon's first year of profitability only arrived in 2003, when net income was a paltry $35 million on revenues of $5.26 billion.[6] Following 15 years of relentless infrastructure spending, building and growth, Amazon's net income went over the $10 billion mark, together with a staggering $258 billion in United States retail sales in 2018. In that year, the company captured 49 percent of the online retail market and had almost single-handedly accelerated the so-called retail apocalypse. By almost any account, Amazon is the sole tech titan that combines the ambition, resources, consumer reach, and trust to become the great disruptor of healthcare.

Many of the long-term investments that Amazon has made speak to the company's obsession with the customer and are also aligned with healthcare. The creation of user experiences such as zero-click (voice-based) and one-click shopping, two-hour shipping, and a constant stream of product recommendations are tangible benefits to their customers. For its customer devotion, the company has been handsomely rewarded. Bezos announced in 2018 that 100 million people globally were signed up with Amazon Prime, with roughly half of the households in the United States in the program.[7,8] Amazon has eroded big box retailing giant Walmart's market share and is competing with Google for product searches on its platform. Amazon's investments in warehouses, information technology, logistics, and customer relationship management set it apart from all other tech giants. The relentless, long-term infrastructure spending by Amazon led to another massive business opportunity—web-accessible computing via access to their server farms scattered in data centers built across the globe. The company offered tools to developers as early as 2004, followed by storage (Amazon S3) and then compute (Amazon EC2) in 2006. By 2020, cloud computing was a $250 billion dollar market with Amazon capturing the lion's share.[9] Among the tech giants, Amazon's AWS owns more than 50 percent of the cloud computing business, with Microsoft's Azure and Google Cloud trailing far behind. The AWS business now accounts for half of Amazon's operating income, and it generated $40 billion in revenue in 2020. On the strength of its cloud computing services, Amazon can effectively reach into any sector of the economy, including healthcare, where IT infrastructure needs and regulatory requirements are sky high.

Amazon has made several explicit moves deeper into healthcare through cloud computing collaborations and services, pharmacy-related acquisitions, and the creation of Amazon employee healthcare clinics. A recent example in the collaboration category was announced in 2019 with the EHR provider Cerner. For its part, Cerner wants to be the software as a service (SaaS) platform in healthcare for its three million providers that log on and access medical records on a daily basis through its system. The company has also seen dollar signs in the mountain of health data that it traffics. The collaboration with Amazon is a transactional win for Cerner and provides Cerner with AWS's machine learning platform, Amazon SageMaker. Cerner has leveraged Amazon's AI technology to create a framework that enables the use of anonymized patient data to build predictive models and algorithms for its customers. AWS now provides a service for doing exactly that with Amazon HealthLake. The concept is to transform all sorts of clinically relevant information into a form that can be analyzed by predictive analytics and machine learning models.

Health data comes in all sorts of formats, from clinically tagged information for medications, medical procedures, and diagnostic codes to unstructured data, such as physician's clinical notes, lab reports, and prescription information. All of these disparate data types are common across healthcare and are not parsable by machine learning tools. Health systems have historically employed people to curate information manually, or they have relied on outdated optical character recognition systems, which are slow and error prone. AI can be used to extract information off of these records using natural language processing algorithms. As the informative data is detected and stored, it can subsequently be indexed and structured in appropriate formats, thereby allowing search queries and use by machine learning algorithms. AWS is approaching insurance companies, healthcare providers, and the pharmaceutical industry to expand Amazon's reach across healthcare by offering new tools and AI infrastructure.

Amazon was situated perfectly for the data explosion and digital transformation in healthcare. High throughput genomics and medical imaging data burgeoned around the same time as AWS was launched. In parallel, digitization of health records created immense storage problems for health systems, where collecting, archiving, and managing clinical data (diagnostics, imaging, health records, mobile app data, and care centers) and nonclinical data from operational systems is essential. Over time, Amazon's sophisticated model for cloud-based management and its array of storage devices has made the company a central player in architectural design of healthcare IT systems.

In 2018, Amazon announced a joint healthcare venture with JPMorgan Chase and Berkshire Hathaway, named Haven, to explore a new model for addressing prescription drug affordability, insurance, and a more efficient delivery of primary care. The new venture never got off the ground and was shut down rather abruptly in 2020. Amid wide-ranging speculation as to why it failed, it appears likely that Amazon's earlier acquisition of Pillpack, the online pharmacy company, factored heavily into Haven's demise. The one billion dollar acquisition of Pillpack, known now as Amazon Pharmacy, is a harbinger of Amazon's ambition in healthcare. The move into prescription drug delivery brings the company into direct competition with entrenched players such as Walgreens, Walmart, and CVS. Another significant move has been the creation of Amazon Care, initially set up as clinics providing primary care and prescription drugs to its Seattle-based employees. The company has partnered with Crossover Health to pilot Crossover's primary-care model at multiple sites near Amazon's fulfillment centers, starting in Dallas, and potentially offering easy access to physicians and other care to more than 100,000 employees. Amazon Care and Amazon Pharmacy appear to be coalescing as the foundation for the retail behemoth to become further integrated into healthcare.

Amazon has assembled additional components to complement its retail-based strategy. The first is a health and wellness product that was built internally by the company to compete with other fitness and health monitors, the Amazon Halo. The screenless device gathers plenty of health data, similar to the Apple Watch and Google's Fitbit. Halo has features for body fat composition, positivity analysis based on speech, sleep, and activity detection and tracking. Since there are already two dozen fitness trackers on the market, the introduction of this product in 2020 seems less about dominating a wearables market and more about long-term strategy. Coupled with Amazon Echo devices at home, the Halo brings another layer of connectivity into retail and recommendations for health and wellness. The company has also made moves on the provider side, grabbing a digital health startup in its acquisition of Health Navigator, which provides tools for clinical decision support and diagnostics.

With its colossal logistics and consumer reach, Amazon is operating from a position of power and using a different playbook than either Google, Apple, or any of the health systems that form the landscape of healthcare today in the United States. The company's retail expertise and customer focus bring the tech giant into healthcare with a different perspective, offering services for comparison shopping and providing convenient access to goods and care, all with a connection to its customer's behavior and lifestyle. It offers convenience at home with Alexa-driven smart devices linked to the world's greatest compute infrastructure and logistics operations. Undoubtedly, Alexa- and cloud-based services will be utilized inside medical centers and doctor's offices as well. Amazon is already connecting hospitals and clinics to medical supplies through its warehouse and transport infrastructure. If Amazon Care expands beyond employee-based programs, it may do so with clinics distributed across the country—probably by an acquisition similar in scale to the purchase of 460 Whole Foods stores—giving them reach into many local communities (and another logistics hub). Amazon is unlikely to venture into drug manufacturing or clinical trials, but elsewhere, the corporation is likely to bring about significant change in healthcare by simply staying on the side of delivery and services.

Echoes of the Final Frontier

The technology sector intersects with healthcare on a variety of levels, from developing microchips for devices to offering services to consumers, insurers, and health systems. The two other tech giants of the current era, Microsoft and Facebook, are not strong contenders as companies that will undertake to transform healthcare or operate a health-based franchise. Although Microsoft is building health data management on top of Azure, the company does not have a social platform or products that serve as a foothold for patients or consumers on any significant scale. Facebook, on the other hand, has the world's largest social platforms—Facebook itself and Instagram—that engage daily with more than a billion people. The company delivers a free service that is part of the fabric of modern life and appears entirely content with a business model that offers other services to business—targeting their users for advertising purposes. Facebook's technology prowess is aimed at building its media empire with better business tools, and healthcare seems an unlikely fit with the company's operating model. Microsoft, along with IBM, Oracle, and others, competes with the other tech titans in business for its share of healthcare IT spending, but it does not have stated goals or a clear path to establishing healthcare as a pillar of its corporate-focused software empire. IBM, Salesforce, and SAP are similarly restricted in scope. Other major tech companies that have dabbled in health and with less-than-ideal profiles for healthcare initiatives are Intel, Qualcomm, and Samsung, computer hardware and mobile device makers with competitive exposure to Apple and Google.

The tech giants of the 21st century have many of the elements to tackle healthcare in transformational ways. Figure 5.2 shows a breakdown of the respective positioning of Apple, Amazon, Google, and Microsoft across healthcare segments. These technology companies have harnessed astonishing compute capabilities through hardware and software, data science, and AI that is unequaled in history. Their proficiency with data is a perfect complement to data-intensive healthcare and the drive toward predictive and preventive medicine. Unlike financial institutions, insurance companies, telecommunications providers, or energy companies that have hundreds of millions of customer accounts, the tech giants combine a similar scale of consumer reach with tangible engagement, offering products and services tied directly to health, wellness, and medicine. It is worth noting that some of the most long-lived incumbents in healthcare, the pharmaceutical companies, only have an indirect relationship with consumers or patients. Technology companies could begin to encroach on the drug companies' turf, with advantages in consumer behavioral targeting, more favorable public trust (at least compared to the drug companies), and luxury brand power.

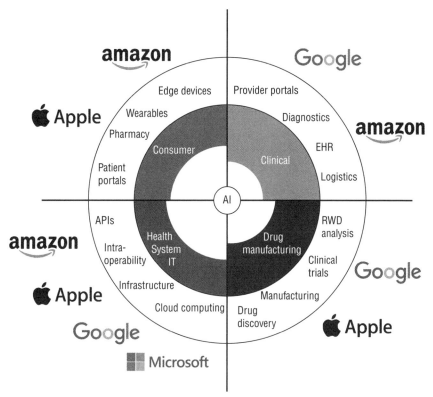

Figure 5.2: Tech giants and healthcare—competitive positioning across the landscape

Technology companies occupy positions in each of the four main quadrants of the healthcare landscape, bringing different technological capabilities and competitive strengths for product offerings and services. In this view, the market is segmented into consumer, clinical, drug manufacturing, and health system IT. The degree to which AI technology is utilized across these market segments is reflected by the size of the inner ring. The set of technologies that can impact healthcare delivery in each segment is shown in the outer circle. The current positioning of the four tech giants—Amazon, Apple, Google, and Microsoft—is shown in the perimeter of the diagram.

The entrance of the tech titans into healthcare will not be about providing healthcare *per se*, but about providing services, products, and efficient delivery, along with enabling new models of care for health systems. Several of these companies are already participating in medicine's expansion into digital health, digital medicine, and digital therapeutics. The engineering mindset at Google, Apple, or Amazon will not lead to a killer software app or a medical tricorder device out of *Star Trek* science fiction. The technology-driven organizations in the end will be most useful in formulating solutions to eliminate barriers to advancing better healthcare for society.

Notes

1. Safavi, KC; Cohen, AB; Ting, DY; et al. Health systems as venture capital investors in digital health: 2011–2019. Npj Digit Med. 2020 Aug 4;3(1):1–5.

2. CB Insights. Global Healthcare Report Q2 2019. 2019.

3. Pichai, S. An Insight, An Idea with Sundar Pinchai. 2020; Davos, Switzerland. Available from: www.youtube.com/watch?v=7sncuRJtWQI.

4. Clement, J. Apple App Store: number of available medical apps as of Q3 2020. Statista. October 2020.

5. Perez, MV; Mahaffey, KW; Hedlin, H; et al. Large-Scale Assessment of a Smartwatch to Identify Atrial Fibrillation. N Engl J Med. 2019 Nov 14;381(20):1909–17.

6. Amazon annual report. 2003. Available from: www.sec.gov/Archives/edgar/data/1018724/000119312504029488/d10k.htm.

7. Reisinger, D. Amazon Prime's numbers (and influence) continue to grow. Fortune. 2020 Jan; Available from: fortune.com/2020/01/16/amazon-prime-subscriptions/.

8. Galloway, S. *The Four: The hidden DNA of Amazon, Apple, Facebook, and Google*. Random House, 2017.

9. Gartner. Forecast: Public Cloud Services, Worldwide, 2018–2020, 2Q20 Update. 2020.

AI-Based Algorithms in Biology and Medicine

The doctor of the future will give no medicine but will instruct his patient in the care of the human frame, in diet, and the cause and prevention of disease.

Thomas Edison, 1903

If we permit flawed machines to make life-changing decisions on our behalf—by allowing them to pinpoint a murder suspect, to diagnose a condition or take over the wheel of a car—we have to think carefully about what happens when things go wrong.

Hannah Fry, 2018

Modern medicine is in the midst of multiple transitions driven by a deeper understanding of human biology and to a greater degree by the rapid introduction of new technology and societal change. The distribution of health knowledge and services over the Internet has led to a shift away from monolithic institutions and toward a more decentralized model, a macro trend that is sweeping up other aspects of life in the 21st century. The iron-fisted grip that the hospitals and physicians once had over information and medical decision-making is changing in part due to these trends and by demands placed on the industry to deliver higher-quality healthcare at lower cost. With any luck, this transformation will lead to dramatic improvements in public health on a global scale.

Against this backdrop, personalized medicine has taken root, impacting drug development, affecting physician-patient interactions, and contributing to the consumerization of healthcare. The concept of personalized medicine emerged from population-scale DNA sequencing studies that revealed the depth and complexity of variation across individual genomes. With this new knowledge, medicine is no longer able to justify a one-size-fits-all approach to health. A necessary transition has led to a more holistic view and awareness of a person's clinical history, socioeconomic and environmental conditions, unique genomic characteristics, and physiological responses to treatment. As a result, healthcare is tackling obstacles to reach a new level of more personalized care. To achieve this, molecular diagnostics and digital health technologies become the main vehicles by which an individual's medical data will be obtained. Whether it is a genomics technology acquiring the molecular signature of a tumor or a wearable sensor measuring heart rate or blood glucose levels, technologies for empowering personalized medicine are leading the way.

The proven or anticipated successes of AI across a wide range of industries stand in stark contrast to what has been achieved and what is expected to occur in medicine's future.[1] Healthcare seems to be the one sacred room where AI is not as welcome—whether by medical practitioners or the general public alike. Society accepts the use of facial recognition by our phones and apps; condones the use of personal information to obtain recommendations on movies, TV shows, and products; expects AI-based detection of fraudulent banking activities; and is becoming better acquainted with chatbots, robotic tellers, and many products that operate by AI. There will soon be self-driving cars, delivery drones, military command centers, and diplomatic negotiations based largely on AI systems. Why not Dr. AI as the family physician, the triage nurse in the emergency room (ER), the lead surgeon in the brain surgery suite, or the entity responsible for monitoring babies in the neonatal intensive care unit? A recent study surveyed a group of 200 students in Boston and New York, asking if they would be comfortable with a medical diagnosis given by a computer with the same level of accuracy as a trained physician.[2] The answer was generally no, and it was not because they didn't trust AI's accuracy. The main reason was that a faceless AI wouldn't be able to understand the uniqueness, or "individualized" disease, that the person possessed. Ironically then, medical AI is at odds with the trend of personalized medicine, a transition already in full swing. Future integration of medical AI into the clinic will need to heed the call for the unique attributes that humans bring to personal health decisions and care.

Physicians harbor the same doubts about AI's role in medical practice, with the additional fear of being replaced like a factory worker competing helplessly against a robot that can work 24/7 and operate with stunning speed and efficiency. Medical training, real-world experience, and human empathy and intelligence are strong elements against any prediction of the doctor's demise in healthcare. The deeply personal nature of caring for someone's health and the trust relationship are guardrails against a techno-utopian view of AI's centrality in healthcare. Much of the current angst about AI in medicine has been around the question of whether machines can ultimately outperform humans in every aspect of medical diagnosis. Tests of expert physicians against trained AI models in diagnostic tasks, such as detecting tuberculosis or cancer in chest X-rays, has demonstrated that these data-driven algorithms, using deep learning, can achieve or even exceed human-level accuracy. The firewall that will likely keep deep learning systems from going too far beyond today's performance is the inability to understand context, to reason. How much common sense and intuition can be built into these algorithms?

Medical training is essentially a long apprenticeship under which physicians become adept at differential diagnoses. The decade's worth of training combined with years of clinical experience produces doctors that are especially good at employing heuristics to make rapid situational assessments. As Eric Topol, the director and cofounder of Scripps Translational Research Institute and author of *Deep Medicine: How Artificial Intelligence Can Make Healthcare Human Again* (Basic Books, 2019), points out, when doctors arrive at a diagnosis in less than five minutes, they are found to have 95 percent accuracy, an uncanny ability. Daniel Kahneman has called this rapid, intuitive-style cognitive skill *system 1 thinking*, which avoids the deliberate, system 2 thinking style needed for careful, reasoned analysis. For clinical situations or presentations with less certainty, doctors must rely on diagnostic reasoning. The clinical diagnostic process involves identifying or determining the etiology of a disease or condition through evaluation of a patient's current symptoms and clinical history, physical examination, consideration of pertinent negatives, and subsequent review of laboratory results or diagnostic imaging. Physicians have access to several diagnostic support systems that augment their capabilities to narrow the search rapidly. A widely used symptom checker algorithm is Isabel (`symptomchecker. isabelhealthcare.com/`). Others are Medscape consult,[3] Simulconsult,[4] and HumanDx. All of these employ search reduction strategies (decision trees, Bayesian networks, and others) to offer up likely diagnoses.

Machine learning tools that have been tested up until recently relied solely on associative inference, and they search for correlations between symptoms and diseases to arrive at diagnostic conclusions. Only a handful of models have been built using causal machine learning, an exciting new approach that incorporates counterfactual inference as part of the diagnostic algorithm. With counterfactual reasoning, an algorithm might seek alternative explanations to explain an outcome, providing the probabilities for "what if?" scenarios. For example, an algorithm could evaluate a counterfactual that considers a bacterial infection as a plausible explanation for a patient's neurological condition given the symptoms. Richens and colleagues designed a study to compare the diagnostic abilities of 44 expert physicians, an associative model and a causal model using counterfactual algorithms. The results demonstrated that their causal model outperformed the associative model and ranked in the top 25 percent of the cohort of expert physicians.[5] The development of a causal machine learning approach for diagnostics is only beginning, and startups like Babylon Health, together with their collaborators in the United Kingdom, are at the forefront.

AI has its algorithmic and critical reasoning work cut out for it in order to take on a central role in mainstream medical diagnosis. Take, for example, the following clinical vignette to illustrate how intuition and experience combine to arrive at a medical diagnosis and decision point in care.

REAL-WORLD MEDICAL DIAGNOSIS: A CHALLENGE FOR AI

Consider a male in his early 50s presenting with symptoms of fatigue, mild cognitive impairment, sleep problems, and irritability. It's a rather vague set of symptoms, but very worrisome. His physician is aware of his prior pattern of alcohol and drug abuse in his 30s and 40s. Were that to have continued, this 52-year-old male would likely now be at risk of cirrhosis of the liver. A quick mental exam confirms that the patient has potential brain dysfunction. Of the hundreds of causes of brain disorders, or encephalopathies, the physician suspects hepatic encephalopathy (HE), a condition resulting from a cirrhotic liver. But the diagnosis of HE is one based on exclusion of other neurological causes. What explains the symptoms? For patients with cirrhosis, precipitating factors could be electrolyte disturbance, medications, gastrointestinal hemorrhage, or infection. The gradual destruction of the liver in cirrhosis leaves the organ incapable of removing ammonia by converting it into urea for secretion by the kidneys. The elevated ammonia (NH_3) in the circulation can be neurotoxic and result in encephalopathy. With that conjecture, the doctor requests a blood test for ammonia (which is not diagnostic for HE, but may guide treatment), and the clinical lab result reveals ammonia levels well into the range of hyperammonemia, a serious condition that can lead to permanent brain damage. At that point the physician admits the patient to the hospital and starts immediate treatment with ammonia scavenging drugs.

In a healthcare system with an AI-based medical ER, would a machine have made the diagnosis? First, conversational AI would need to be good enough to get a patient–doctor dialogue going to tease out whether the patient truly was impaired, maybe drunk, or some other in-between state. The AI agent would then need to access clinical records and make the connection to cirrhosis, direct the patient to a blood draw center, and then assess the clinical lab report. How would the AI exclude other causes? Failure to act quickly with this condition could lead to permanent brain damage, coma, or death. The criticality of such diagnoses is why physicians, envisioning a repeat of these types of scenarios over and over again for millions of patients, have not bought into the superhuman AI medical scenario of the techno-utopians.

Regardless of these cautionary examples, AI is a core technology for the future of healthcare. The acceleration of AI capabilities, especially deep learning's godlike performance around image recognition tasks, has led journalists, AI experts, and a few authors of bestselling books to predict that physicians may one day be replaced by AI systems. Applications of artificial intelligence in biology and medicine have a natural home in diagnostics where pattern recognition and classification algorithms can wield their superpowers. With AI, the frightening part to many is not its use in behind-the-scenes testing and image-based diagnostics, but in medical diagnosis.

The research community is setting a fast and furious pace to test and improve algorithms across the entire range of applications, as listed in Table 6.1. Just a few years ago, the main data type used as a testbed for AI in medicine was largely confined to X-rays, a digital native image format that was amenable to computer vision algorithms. Nowadays, the data being fed to AI models essentially comes from all corners of the medical and molecular diagnostic world.

Table 6.1: Current Applications of AI-Based Algorithms in Medicine

DATA TYPE	CLINICAL USE	THERAPEUTIC AREA	PREDICTION/ MODEL
X-rays			
X-rays	Therapy guidance	Osteoarthritis	Pain sensitivity[6]
X-rays	Diagnosis	Orthopedics	Bone fracture detection[7]
Mammograms	Screening	Cancer	Breast cancer[8]
CT scans	Diagnosis	Neurology	Diagnosing neurological disease[9]
MRI images			
Brain MRI	Prognosis	Cancer (Glioma)	Survival prediction[10]
Brain MRI	Diagnosis	Multiple sclerosis	Identifying predictors from CNNs[11]
Brain MRI	Prognosis	Multiple sclerosis	Predicting disease course[12]
Video			
Colonoscopy images	Diagnosis	Gastrointestinal	Polyp detection[13]
Pathology slides			
H&E WSI	Diagnosis	Liver	Tumor classification[14]
Biopsy	Diagnosis	Cancer (DLBCL)	Tumor Classification[15]
Biopsy	Diagnosis	Prostate cancer	Grading biopsies[16,17]
H&E sections	Diagnosis	Neurology	Neurodegenerative disease classification[18]
Biopsy WSI	Diagnosis	Breast Cancer	Classification with Ki67 marker[19]
Electrocardiograms			
12-lead ECGs	Diagnosis	Cardiology	Rhythm classification[20]

DATA TYPE	CLINICAL USE	THERAPEUTIC AREA	PREDICTION/ MODEL
EHR data			
EHR data	Patient stratification	NA	Disease subtyping[21]
EHR data	Prognosis	Nephrology	Post-operative outcomes[22]
Clinical vignettes	Medical diagnosis	All	Diagnostic accuracy[5]
Molecular data			
DNA methylation	Diagnosis	Cancer	Cancer classification[23,24]
DNA sequence	Diagnosis	Cancer	Cancer classification[25]
Multimodal data			
EHR and imaging	Diagnosis	Pulmonary vascular disease	Detection of pulmonary embolisms[26]
Pathology slides and clinicogenomic data	Therapy selection	Cancer (lung)	Patient stratification for immunotherapy[27]
Pathology slides and molecular marker data	Diagnosis	Cancer	Tumor classification and mutation detection[28,29]

Recognizing the Faces of Cancer

Cell and tissue, shell and bone, leaf and flower, are so many portions of matter, and it is in obedience to the laws of physics that their particles have been moved, moulded and conformed. They are no exceptions to the rule that God always geometrizes. Their problems of form are in the first instance mathematical problems, their problems of growth are essentially physical problems, and the morphologist is, ipso facto, a student of physical science.

D'Arcy Wentworth Thompson, *On Growth and Form*, 1917

The genetic changes that lead to uncontrolled cell growth and neoplastic transformation produce a constellation of signatures at every level of biological organization in cancer. Over an organism's lifespan, environmental damage and metabolic insults leave scars on DNA, such as changes in methylation status at single bases that are the smallest of molecular alterations that can be evidenced across the genome. These epigenetic changes, often occurring a decade prior to the formation of neoplasms, are recognized now, as harbingers of cancer. Eventually, DNA mutations accumulate, allowing oncogenic drivers to orchestrate cancerous genetic programs that first reveal themselves as alterations in RNA copy numbers and a remodeled transcriptome. Chromatin remodeling and chromosome-level modifications, such as telomere DNA maintenance, subsequently arise through triggered mechanisms that evade normal growth control. The hallmarks of cancerous growth are mostly clearly seen as morphological changes to cell shape and size, and an ongoing cellular reorganization leading to a characteristic tumor microenvironment in tissues. In one of the largest studies of its kind, classification of 10,000 tumor samples based on genomic and transcriptomic molecular data revealed that cancer types are tightly linked to histology and tissue type, organized primarily by cell-of-origin.[30] This linkage of molecular entities to morphological features provides an entryway for bringing in AI, and computer vision in particular, for their pattern recognition capabilities applied to cancer detection and classification tasks.

In oncology, diagnostic imaging is often the first step in screening, and it provides definitive diagnoses for many cancers, including those originating in brain, lung, or breast tissue. Patients presenting with signs of brain malignancy will first be examined by MRI to image and detect the presence and location of a tumor. A more precise diagnosis and treatment plan can be aided by a biopsy to examine the cells, if surgery is available as an option. A follow-up PET or CT-PET scan can provide additional information on tumor progression and potential treatments. Finally, molecular diagnostic testing might be employed to pinpoint markers that are of therapeutic or prognostic value. In breast cancer, imaging plays a central role in screening and diagnosis, with both steps performed by mammography. Diagnosis is augmented or confirmed in some cases with additional imaging by ultrasound or MRI. A positive test leads to biopsy to evaluate whether the growth is invasive or noninvasive and as a starting point for revealing a host of prognostic factors. Histopathology done on tissue biopsies is used to determine status (presence/absence) of important biomarkers, mainly HER-2, estrogen receptor, and progesterone receptor. Additional genomic testing can be conducted with focused gene panels such as OncotypeDx from Genomic Health. In advanced, metastatic, refractory, or recurrent cancer, molecular testing will be used to guide therapy further in combination with blood tests to assess liver and kidney function, immune function (complete blood count or CBC), and the presence of certain viruses that can impact treatment regimens (Hepatitis B or C, or SARS-CoV-2). The diagnostic process thus runs the gamut from gross observation taken during a physical exam to high resolution, image-based detection of abnormal growths and cellular level tumor classification. Add to this a detailed molecular characterization, which provides the rationale for personalized intervention strategies and aids in monitoring the response to therapy and disease recurrence.

Deep learning has shown its greatest promise in medicine as a tool for oncologists, where intense evaluation has begun on how AI systems can be used at each of the critical steps of the clinical pathway. One of the first hurdles to overcome in integrating AI diagnostic support tools into the clinic is the proof that a doctor's decision-making abilities are improved—and not unintentionally hindered—by AI. Andrew Ng, Jeanne Shen, and colleagues at Stanford completed an important study along these lines in 2019 (published in 2020) by asking, how would an AI assistant impact diagnostic performance of pathologists with varying levels of expertise?[14] What the team found by employing a clever study design was revealing. The deep learning algorithm they had developed was powerful but possibly too persuasive. The AI decision support technology was designed to aid pathologists in distinguishing hepatocellular carcinoma (HCC) from cholangiocarcinoma (CC) from standard, whole-slide images of hematoxylin and eosin-stained tissue sections of primary liver tumors. The model by itself achieved a mean accuracy of 0.842 on an independent test set of 80 slides. In the test, four groups of pathologists were tasked to evaluate the 80 slides, with or without assistance. In the crossover design, each pathologist reviewed half of the 80 slides without assistance and half with assistance. After a two-week washout period (designed to eliminate short-term memory of the slides/data), the assistance order was reversed for each slide.

The study emulated a digital pathologist's workflow. Following visual inspection and evaluation, the pathologist selected a region on the slide and then uploaded a scanned image into a cloud-based system that ran the prediction algorithm. Probabilities were returned for predictions of either HCC or CC, and pathologists could use the information to make a final call while in assistance mode. The use of the AI assistant improved the level of accuracy for nine of the eleven pathologists in the study in three of the four groups: those who were either GI subspecialists, non-GI specialists, or medical trainees. When the researchers looked more closely at the data, they found that when the AI model got the call right, a pathologist had 4.3 times the odds of coming to a correct diagnosis compared with those pathologists not using assistance. What was startling, though, was that when the AI predictions were wrong, the support tool left pathologists with less than one-third the odds of making the right call. The dilemma then in using AI tools in a diagnostic setting is the concern that it will cause doctors to override their good judgment as a result of trusting a persuasive intelligent machine. See Figure 6.1.

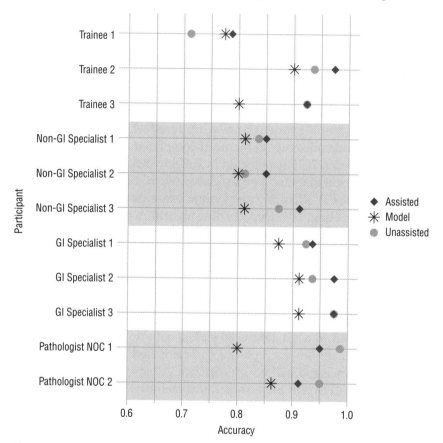

Figure 6.1: Impact of assistance on individual pathologist diagnostic performance

The average diagnostic accuracy (across the set of 80 experiment WSI) for each pathologist is plotted as follows: gray circle (unassisted) = accuracy of the unassisted pathologist, star (model) = accuracy of the model alone (based on pathologist selected input patches), and dark gray diamond (assisted) = accuracy of the pathologist with model assistance. Figure and legend reproduced from Kiani, A., Uyumazturk, B., Rajpurkar, P. et al. Impact of a deep learning assistant on the histopathologic classification of liver cancer. *npj Digit. Med.* 3, 23 (2020). doi.org/10.1038/s41746-020-0232-8. Under creative commons license: creativecommons.org/licenses/by/4.0/.

Another daunting task for oncologists in particular is predicting patients' life expectancy after it is determined from diagnostic tests that metastatic cancer is present. The ability to deliver a prognosis accurately is important for anticipating care needs and, most importantly, for patient and family when making life decisions around the awful prospect of terminal cancer. AI prognostic models have only begun to be developed and tested, with clinical researchers hoping to improve upon earlier statistical models that were less accurate than physicians in estimating length of survival.

An important first step in AI prognostic model development emerged from a large-scale study that was conducted by Michael Gensheimer and colleagues.[31] The team began by training a machine learning model with a neural network architecture on EMR data from 14,600 metastatic cancer patients taken between 2008 to 2020 in the Epic system at Stanford Health Care System. Out of the records, they fed the network with a 3,813 × 1 feature vector—this was from predictor variables extracted from care provider notes and radiology reports (n=2,562 words as predictors; others included laboratory values, n=319, vital signs, n=4, and so on). For the study, 899 metastatic cancer patients were enrolled between 2015–2016. While in the hospital, radiation oncologists estimated patients' life expectancy and the research team compared predicted length of survival by these physicians, the machine learning model, and a traditional model. Using a standard metric of performance known as *area under the receiver operating characteristic curve*, abbreviated typically as *AUC* (or AUROC), their analysis revealed that the machine learning algorithm (0.81 AUC) outperformed both the physicians (0.72) and a traditional model (0.68).

Given the complexities surrounding the study variables and physicians' considerable knowledge, it is a remarkable feat to demonstrate that an algorithm, fed a collection of EMR data, can return a consistently better prediction in many cases than experienced medical professionals. The question then arises as to whether such predictions produced by a clinical decision support tool, if inaccurate, would harm patients or make the physicians prognosis worse, as in the case of liver cancer (outlined earlier). It is also unclear whether the prognostic performance in metastatic cancer seen in a top academic hospital environment would apply elsewhere (for example, in settings with less reliable EHR records and different standards of care). Further algorithm development and appraisal of AI-based decision support frameworks in prospective clinical trials will be needed to answer these important questions.

Tumor Classification Using Deep Learning with Genomic Features

The examples in the previous section highlight both the progress and the challenges in clinical implementation of AI technology for oncology. The use of imaging-based techniques based on CNNs (as outlined in Chapter 2) has dominated the field of AI-driven tumor classification. Some of the most exciting new terrain is found where AI is being used to predict tumor types by molecular classification alone. Inroads are being made by letting artificial neural networks work with DNA methylation data, transcriptome signatures, and genetic mutations as features. An integrative diagnostic approach is envisioned that combines both histopathological and molecular data to detect and classify tumors more accurately. Figure 6.2 shows these types of research and clinical applications and how they intersect with AI models and architectures.

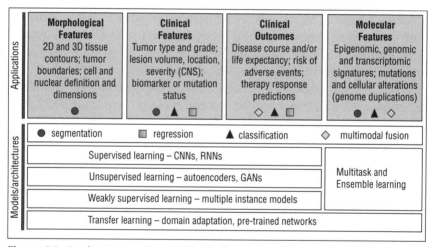

Figure 6.2: Applications and models for AI-driven algorithms in medicine

Deep learning approaches have led to advances in applications to derive medically and biologically important features from a variety of data sources (as outlined in Table 6.1). The top panel summarizes four main applications. The bottom panel provides the corresponding models and architectures used for algorithm development in the applications. Supervised learning with CNN architectures has dominated development of segmentation and classification algorithms for tasks related to detecting cell and nuclear boundaries, tumor and disease state classification, and representation learning from high dimensional data found in molecular profiling. Transfer and multitask learning have been important for development of new models using biological data. Multimodal fusion models are used in applications requiring integration across data modalities.

Epigenomic investigations have benefitted from algorithmic approaches that are able to associate DNA methylation patterns with tumor types. A compelling study was completed by an international team of researchers with genome-wide DNA methylation arrays that showcases the power of AI to uncover epigenetic signatures for a wide range of brain tumors.[23] To begin the work, DNA was extracted from a set of 2,801 tumors and methylation profiling was completed on processed DNA samples with Illumina instrumentation. The team's aim was to survey histological tumor types comprehensively to define a reference cohort. This was accomplished by using both unsupervised and supervised machine learning to recognize patterns and features for subsequent classification.

The results were impressive. The algorithm found 82 methylation-based classes, uncovering many novel, unsuspected, or previously unidentified tumor types, and it was able to refine known tumor subtypes. On the basis of the brain tumor methylome alone, the AI technology was able to capture all of the histological classes (embryonal, glioblastoma, glio-neuronal, mesenchymal, and so on) and left no doubt as to the diagnostic potential of an integrated histologic and epigenomic approach. To evaluate the clinical value and concordance with pathologists, the method was performed on 1,104 test cases. For 60 percent of the cases, the machine-derived classification agreed with expert humans. In an additional 16 percent of cases, the difference was a matter of the AI system pinpointing a subclass that was not apparent to the pathologist or recognized in current brain tumor classification schemes. Diagnostic concordance was thus in the neighborhood of 75 percent. There was disagreement in 12.5 percent of cases. But when these samples were extensively analyzed by other genomic methods, 92.5 percent of the machine-based classifications were confirmed to be correct.

The results from this large, retrospective study have clear implications for how classification algorithms and DNA methylation profiling could be used in clinical trials and eventually in clinical settings. Accurate subtyping would allow standardization in clinical trials, which are typically multicenter with different pathologists identifying and grading tumors. A prospective clinical study is underway in Germany to refine and validate the AI-based diagnostic tool for classifying childhood brain tumors. This epigenetic-focused study reaffirms that adding molecular characterization, which was already done for some brain tumor assessments, would add value to clinical decision-making. However, the implementation of a new and rather complex workflow is unlikely to be easily accomplished due to cost and data integration details, nor are pathologists ready to disrupt their standard microscopic examination workflow.

A second genomic-based approach to tumor classification that is conceptually similar to the methylation profiling strategy was conceived by a team using whole genome sequencing data and reported in 2020.[25] The clinical question that the researchers addressed was whether a collection of structural or sequence variants distributed throughout the genome—a molecular signature—could serve as an accurate proxy for tumor type. Since driver mutations are only a very minor fraction of the expected mutations found in tumors, the underlying signature would be composed mainly of passenger mutations. From a set of 2,606 tumors representing 24 types of cancer, whole genome sequencing data was obtained and used to train a series of machine learning models for classification. The high-performance model was, as expected, a deep learning neural network, which demonstrated 91 percent accuracy on the validation set. When tested on an independent set, classification accuracy was 88 percent for primary tumors and 83 percent for metastatic tumors. The investigators also looked under the hood to see where the classification power was coming from, and the most informative features were those from single nucleotide variants and small indels that comprised passenger mutations; driver mutations and large structural variants had little effect on aiding performance. The remarkable findings with DNA sequence variants and tumor type reinforce the idea that patterns of somatic passenger mutations and DNA methylation signatures encode the "cell of origin" state that is the strongest determinant of tumor type and clinical manifestation.

Genomic tests and molecular profiling are, of course, not new to medicine, oncology, or anatomic pathologists. The primary reasons for their slow adoption are cost per test and slow turnaround times for results back to physicians. The use of molecular diagnostics is most often sought to personalize treatments based on specific markers, such as with the identification of oncogenic mutations in the EGFR gene for lung cancer and in BRAF for certain melanomas. What if the microscope could recognize these types of mutations from the cellular morphology alone? And better yet, could AI be used to infer genotypes from images, a reversal of the more common genotype to phenotype explorations common in biology. This concept has precedence in colorectal cancer, where microsatellite instability markers are well-correlated with morphological changes and some tests are approved for diagnostic use.[32,33] In 2020, several important, paradigm-shifting studies revealed the extraordinary capability of deep learning–based computer vision algorithms to be trained and built for multipurpose diagnostics, potentially unlocking image-based genetic testing and fully integrative approaches as game-changing alternatives for the future.

The ease of testing algorithmic approaches and network architectures on available cancer imaging and molecular profiling datasets has accelerated explorations across all cellular mechanisms that impact cancer biology. The first studies of this type evaluated algorithms trained to predict a narrow, or single class of, molecular change from histology slides.[34] Among the swarm of investigations underway to use morphological features to predict molecular status, detecting DNA mutations in cancer has consistently shown great promise. In one of the first studies of its kind, a group led by researchers from New York University used a collection of images from the NCI Genomic Data Commons to predict mutations in non-small cell lung cancer using image data alone as input to the model.[35] The CNN, built with Google's Inception v3 architecture,[36] was adapted for multitask classification and was able to determine the mutational status for a set of 6 genes (out of 10) with reasonable accuracy. The best performance was for STK11 (0.85), and the new algorithm notably could predict status for other important cancer drivers such as TP53, KRAS, and EGFR in lung adenocarcinoma.

The results from this work were among the first definitive demonstrations that deep learning CNNs could be used to work backwards from images to obtain genetic information useful for histopathologic diagnoses. The task was demanding, as tumor tissue is highly heterogeneous and allele frequencies vary from gene to gene across tumor types (in this case lung adenocarcinoma [LUAD] and lung squamous cell carcinoma [LUSC]). Apart from the technical achievement, it revealed that sets of morphological features can be associated with individual genes whose encoded protein products are, by themselves, not obviously structural components. Indeed, most of these identified genes encode components of signal transduction cascades whose effects on morphology are achieved by regulating hundreds of genes acting in concert on cellular growth.

The team at NYU also built a deep learning classifier to distinguish among lung cancer types and normal tissue. The system was nearly perfect (AUC ~0.99) for classifying normal versus tumor (see also Chapter 2, Table 2.2, for similar work with LUAD and LUSC), illustrating the power of *transfer learning* paradigms for biological data analysis. A common practice is to pretrain models on ImageNet or use network parameters obtained from parameter sets with best performance on the dataset. By doing this, features learned from natural images, or others, can be transferred to another domain, such as biology. The underlying premise is that statistical strengths are shared across tasks. Regardless of the domain where the image has come from, lower levels of these networks will capture similar features (edges, blobs, contours) that are not specific to the training data and the application, but instead recur in perceptual tasks in general. An important practical outcome of transfer learning is that training demands are lessened, requiring fewer images to fine-tune the model for the task of interest. Another benefit of the Inception architecture is that it is well-suited for tissue imaging, as it has convolution modules at various resolutions, allowing for capture of features at different scales.

Given the success in predicting the presence of molecular details from histology slides, researchers have begun investigating the ability to associate morphometric features with more complex transcriptomic or epigenomic signatures or profiles. These modalities have proven equally accessible via deep learning strategies. Extensive sets of RNAseq (transcriptomic) data are available to the research community through the GTEx Portal (`www.gtexportal.org/home/`) or GDC Data Portal (`portal .gdc.cancer.gov/repository`). Information from slides and transcriptomic profiles can be used to classify tissues, cancer subtypes, and other disease states. The beauty of these integrative strategies is that interpretability is high: features can be directly tied to genes, with further understanding derived from common features shared across genes within the same tissue or across cell states such as cancer.

Epigenomic influences on cellular morphology were a logical next step to unravel with AI techniques. Were the effects of DNA methylation or histone modification on cell morphology strong or weak, and where did the signals reside? A study run on cohorts from TCGA where both methylation and imaging data was available for glioma (342 patients) and renal cell carcinoma (326 patients) provided a clear roadmap of how to approach these questions.[37] As expected from earlier large-scale DNA methylation profiling work[23], machine learning methods can be used to detect the association of DNA methylation profiles with features extracted from whole slide images. The process first required methylation states to be inferred from the profiling array on a gene-by-gene basis. The array of DNA methylation status (DM) values is then placed in a gene (columns) by patient (rows) matrix. Morphometric features are extracted by conventional image processing techniques. Similar to the lung study, a *multitask learning* framework was employed, where each gene is a "task" and each features a variable for the multivariate classification problem. The researchers applied several machine learning models to predict DM values by using the features.

In July 2020, two independent studies[28,29] published simultaneously in *Nature Cancer* broke new ground in digital pathology, showing that a single deep learning algorithm can be trained to predict a wide range of molecular alterations across many cancer types (referred to as *pan-cancer analysis*). Yu Fu and colleagues used an Inception V4 architecture[38] and transfer learning to build their pan-cancer computational histopathology tool, PC-CHiP. The strategy leveraged a set of 1,536 features to classify 28 cancer types and predict survival, genetic mutations, and gene expression profiles. Following a different technical path, an international team tackled the pan-cancer analysis problem with multitask learning and an end-to-end computational workflow to predict four sets of molecular features from histology images. The results from these clinical studies were stunning, painting in technicolor the ability of these deep learning systems not only to work across a wide array of cancer types but to characterize very different types of molecular alterations from slides alone. The two studies, whose results were driven primarily by deep learning, provide further evidence of the link between the morphology of the tumor and the underlying molecular features that exists in all cancer types.

Several biotechnology startups have entered the challenging arena of computational medicine, building clinical decision support systems or AI-assisted instrumentation for oncology. Breast cancer appears to be the top application for technologies on the leading edge. Clinical decision support systems for therapy guidance and selection have been one focus, aimed at augmenting physicians' decision-making during consultations. A few teams have built predictive models from multiple modalities and also base predictions in part on a thorough metabolic pathway analysis. Given a patient tumor's "omics" profile, these systems can pinpoint cellular defects or metabolic pathways that might render first-line therapies less effective, resulting in lost time and slower arrival at more effective therapeutic regimens. The algorithms can generate recommendations for alternative therapies to standard of care along with tumor growth predictions under various chemotherapeutic scenarios.

A company that has been active in developing metabolic pathway and AI clinical decision support tools is the Cellworks Group. From a recent abstract presented at the American Society of Clinical Oncology's annual meeting (2020), "The Cellworks Singula™ report predicts response for physician prescribed therapies (PPT) using the novel Cellworks Omics Biology Model (CBM) to simulate downstream molecular effects of cell signaling, drugs, and radiation on patient-specific *in silico* diseased cells."[39]

Another innovative company that has tumor virtualization technology is SimBioSys, Inc., founded out of research at the University of Illinois at Urbana – Champaign. The SimBioSys team has constructed a computational oncology workflow and physician-oriented software platform with an initial focus on breast cancer. Their algorithmic approach is physics-based, starting from the laws that govern the chemical, physical, and mechanical interactions within and between cells. This unique approach provides a comprehensive prediction of response to therapy, thereby enabling discovery of novel insights as well as optimization of clinical trials. Figure 6.3 illustrates an overview of the platform.

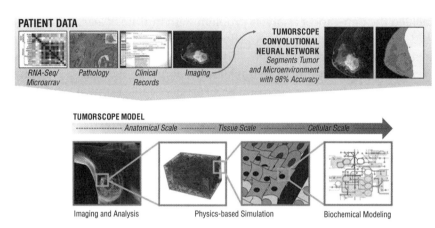

Figure 6.3: Computational oncology platform from SimBioSys

Top panel: Overview of elements comprising the computational workflow. **Bottom panel:** SimBioSys TumorScope® is a proprietary biophysical modeling software platform to assess the efficacy of drugs on a personalized virtual tumor for a patient through computational simulations. The simulation engine combines all major hallmarks of cancer variation to understand cancer comprehensively, assessing the impact of drug delivery, drug sensitivity, metabolism, and spatial heterogeneity on tumor biology.

Figure courtesy of Tushar Pandey, CEO, SimBioSys, Inc.

For these startups, implementing clinical AI faces a battle with payors and providers as a result of the long-established practice of adherence to standard of care and first-line therapies. Encouraging a physician to consider using a clinical decision support tool is not too difficult; placing physicians in a situation where they have to decide whether to order an alternative treatment based on an AI system recommendation that is contrary to conventional standard of care is quite another. Clinical decision support systems will likely need prospective clinical trials to prove their accuracy and utility in offering higher-performing options for chemotherapy regimens and drugs for certain cancers.

Data science and AI are also critical pieces for liquid biopsy companies developing early cancer screening and mutation detection capabilities out of blood DNA samples. Grail (acquired by Illumina in 2020), Thrive

Earlier Detection (acquired by Exact Sciences in 2020), and Freenome use signals from deep DNA sequencing and DNA methylation analysis together with logistic regression methods to spot mutations.

AI for Diseases of the Nervous System: Seeing and Changing the Brain

The brain is a far more open system than we ever imagined, and nature has gone very far to help us perceive and take in the world around us. It has given us a brain that survives in a changing world by changing itself.

Norman Doidge, in *The Brain That Changes Itself*

The constellation of neurological diseases, acute brain injuries, and psychiatric disorders that can potentially arise over a human's lifespan present both extraordinary challenges and groundbreaking opportunities for the use of AI-based algorithms in neurology. Nervous system diseases share at a superficial level one characteristic of cancer; namely, they constitute a collection of diverse conditions with poorly understood or undiscovered disease subtypes, as for example with the many types of dementia. Disease heterogeneity across individuals makes these conditions difficult to diagnose, even with sophisticated imaging modalities, an array of highly sensitive biomarkers, and experts armed with well-designed clinical assessment tools. As with cancer, an individual's genetics and environment play prominent roles in disease manifestation. Neurodevelopmental disorders such as autism and psychiatric diseases, including schizophrenia, take root during nervous system formation and brain development, which is not completed until a person has entered early adulthood. The plasticity of the brain is a major feature driving development, and this capacity is also retained, to a lesser degree, after full maturation. Thus, the brain can be impacted by traumatic experience early or late in life, and its inherent plasticity affords opportunities for both harm and healing, reprogramming, and routes to recovery and improved health. AI that can be directed to interact with the brain's changeable nature—for movement, sensory perception, or other functions—would give clinicians another tool for tackling an array of neurological conditions.

AI is in its infancy with regard to interacting with brain function, unless one considers all the algorithms designed for retaining attention to social platforms or altering shopping behavior. A few companies, such as Modality, have engineered conversational AI systems that integrate speech and facial responses for monitoring neurologic function and mental condition. The AI tool being developed by Modality may generate data of sufficient quality for clinicians and researchers to enable detection of response to therapies or altered living conditions. Modality has a system that will probably be replicated by the ambient computing devices available in home or clinics and offered by Amazon, Google, Apple, and Microsoft. Regardless of the route by which conversational AI systems reach patients and connect to healthcare systems, there is tremendous room for human-AI interactions to improve quality of life, especially for those with neurological impairments and temporary disabilities.

Neurology, like oncology, is highly dependent on imaging technology for accurate medical diagnosis. As with other areas of medicine, AI is poised to have a great impact on brain imaging tasks and image interpretation. Across neurology, examples are proliferating for AI in automating feature extraction, predicting disease classes and outcomes, and improving neurological assessments. From a clinician's standpoint, these tools will provide needed automation of feature extraction from images and serve as a critical clinical decision support tool for patient treatment and disease monitoring tasks.

Progress in the use of AI in neurology is found by tracking the source of images. In Alzheimer's disease research, the equivalent to cancer's massive TCGA datasets originates in the Alzheimer's Disease Neuroimaging Initiative (ADNI). The ongoing longitudinal study has a collection of MRI and PET brain scans of more than 3,000 participants from several cohorts that is shared through the USC Laboratory of Neuro Imaging's Image and Data Archive.[40] The laboratory's archive houses the world's most prominent collection of neuroscience research data, currently at 141 studies with biomedical information on more than 55,000 subjects in diseases ranging from autism, Alzheimer's, Parkinson's, and Huntington's to traumatic brain injury and the human brain connectome project. The availability of longitudinal imaging data for Alzheimer's disease provides researchers with abundant opportunities to test new AI-based models against current diagnostic protocols. Early research with ADNI study data prior to deep learning strategies with image information indicated that various machine learning approaches were suitable for forecasting some aspects of early Alzheimer's disease pathology.[41–43]

A more recent advance resulting from work by scientists at UCSF has shown that a deep learning model trained on PET data can detect changes associated with the early, presymptomatic phase of Alzheimer's disease up to six years prior to disease onset.[44] One fascinating aspect of the study was the presentation of saliency maps of the imaged brain, derived in an attempt to visualize how the algorithm was making its decisions to categorize individuals into those with either full-blown Alzheimer's disease, mild cognitive impairment (MCI), or no disease. The systematic analysis of saliency maps could one day unveil a new imaging biomarker that neurologists can use during clinical assessments. Unfortunately, there were tantalizing traces of regional hotspots of neuropathology in this study, but no clear signal, suggesting that the algorithm was utilizing information taken across the entire brain into consideration when making the predictions.

The PET-based pilot study combined with AI to produce some of the first results of using functional imaging, rather than structural MRI, in predicting neurological disease. The investigators pointed out several caveats in interpreting the AI's performance as a *bona fide* clinical tool, but the future is bright for new study designs and improved algorithms to help doctors catch the onset of Alzheimer's disease where preventive strategies will wield their greatest impact. Other efforts are underway to test and validate AI algorithms with MRI data, a cheaper alternative that in the long run would be more amenable to healthcare adoption.

Diagnostics and therapeutics development for Alzheimer's and other neurodegenerative diseases have had to grapple with extremely complex disease biology in the brain and the often-byzantine clinical presentation of neurological symptoms. Over the past several decades, a strong consensus has been built around the idea that the neurodegenerative process is driven by pathologically aggregating proteins, fueling diseases known collectively as *proteinopathies*. These disease-associated aggregates are familiar in the case of Alzheimer's disease, with toxic amyloid-beta found in neuritic plaques and the microtubule-associated protein tau present in neurofibrillary tangles. Parkinson's disease, multiple system atrophy, and Lewy body disease are characterized by α-synuclein inclusions, whereas brains of frontotemporal dementia patients have tau or TDP-43 inclusions. In ALS, TDP-43 cytoplasmic aggregates appear frequently in upper motor neurons in the brain. Prion protein aggregates are hallmarks of Creutzfeldt-Jakob disease. For many of these diseases, it is now clear that aggregating proteins also interact, and brains of diseased individuals typically harbor two or more distinct aggregate types in visible extracellular plaques and intracellular inclusions.

Clues as to the nature of neurodegenerative disease pathology came in the 1990s from DNA sequencing and genetics. Inherited mutations in some of the genes encoding these proteins lead to the generation of toxic, aggregating species, such as with amyloid-beta, or induce protein misfolding and subsequent aggregation. Mutations in the human TDP-43 gene are not found in the vast majority of ALS patients, but the protein is aberrantly modified. The nature of the disease-initiating steps in proteinopathies and the role that the aggregates play in the ultimate destruction of the affected brain regions are a matter of intense debate and research.

The ability of neurologists to diagnose these clinically overlapping disorders accurately is nearly impossible, as patients often meet criteria for multiple diseases and current tools are focused on evaluating only one or two biomarkers. To begin to tackle this diagnostic challenge, a study that broke new ground utilized a set of statistical models to analyze co-pathology among several of the aggregating pathogenic proteins mentioned earlier, in addition to neuronal loss, gliosis, and angiopathy across 15 brain regions in post-mortem brain samples.[18] In total, 98 features were used to categorize 895 samples from patients diagnosed with neurodegenerative disease by unsupervised clustering tools, which defined six core clusters. The results produced a significant reshuffling of diagnoses, clustering individuals into proteinopathy families that correspond to new transdiagnostic categories. The clusters themselves were nonoverlapping, organized around one or more aggregating proteins: tau (cluster 1), amyloid-beta and tau (cluster 2), TDP-43 (cluster 3), α-synuclein (cluster 4), amyloid-beta and α-synuclein (cluster 5), and cluster 6, characterized by low cerebral pathology and no prominent signal for these pathogenic proteins.

What is striking to nonexperts is that a primary diagnosis of Alzheimer's disease is nowhere near the end of the story of what might be happening in the diseased brain. Individuals with an Alzheimer's diagnosis in this large study group were found across all clusters, or transdiagnostic categories, and more than a dozen other neurodegenerative diseases lurked behind the scenes. A few of these, such as Lewy body disease, a common form of dementia, are concomitant with Alzheimer's and were reported as a secondary diagnosis in a significant subset of individuals. The clustering revealed these shared connections by using information from the molecular data inferred from the histopathology alone, and it suggests that neurological patients may suffer from additional, latent pathologies that a neurologist cannot diagnose. On a practical level, the researchers were able to train a logistic regression model that accurately predicted membership into these six categories, using cognitive test scores combined with biomarker levels and genotyping. This type of data science-driven approach could work in identifying disease subtypes more broadly, outside of the world of neurology.

The trove of brain imaging data that has accumulated in the study of Alzheimer's, specifically from ADNI, has been beneficial for other areas of neurological research with the application of transfer learning. A great example of this was shown for building an AI-based diagnostic for multiple sclerosis (MS). Using a set of data from 921 participants in ADNI to pretrain a deep learning model, Fabian Eitel and colleagues were able to classify people with MS versus healthy volunteers at 87 percent accuracy.[11] A much more challenging task in MS is to predict the trajectory of disease. One such study was initiated over a five-year timespan with 724 patients from the Comprehensive Longitudinal Investigation in MS at Brigham and Women's Hospital (CLIMB cohort) and validated with 400 patients from the EPIC dataset at UCSF.[12] The classification task here was to forecast from clinical and MRI data whether the disease condition would worsen at five years, based on information taken at the two-year mark. Here the team built a series of models with a subset based on conventional machine learning models (SVM, logistic regression, and random forest) and a set using ensemble learning (XGBoost, LightGBM, and Meta-L). Model performance was slightly better with the ensemble approaches, with AUCs ranging from 0.79 to 0.83. Although far from a perfect predictor, these results are impressive in view of studies conducted on the same study cohorts that found significant differences in disease course and demographics that limit the strength of correlations expected.[45]

Deep learning approaches for image classification have shown excellent performance for other areas of neurology, including epilepsy, stroke, and other acute neurologic events, such as hemorrhage and cranial fractures. Many neuroradiology scans are done in the clinic to obtain volumetric (3D) data, a more challenging task for AI models to train against. A team led by Erik Oermann built a 3D CNN model for detecting acute neurologic events using 37,236 CT scans, which were annotated with nearly 100,000 radiology reports. The researchers set up a randomized, prospective clinical trial to test algorithmic performance versus radiologists in an AI neuroradiology workflow that could be used to triage patients in a hospital/ER setting. The trained 3D CNN exceeded expert radiologists' performance in ability to prioritize the most urgent cases. Not surprisingly, the AI was 150-fold faster. The study was one of the finer examples of rigorous testing of a deep learning–based workflow in a simulated clinical environment.

Regulatory Approval and Clinical Implementation: Twin Challenges for AI-Based Algorithms in Medicine

An excellent measure of the innovative impact of AI on medicine is the growth rate in approvals of AI/ML-based medical devices in the United States and Europe. A comprehensive study surveying the use of machine learning in medical device software indicated that the first series of approvals were issued in 2015.[46] The software as a medical device (SaMD) category had predecessors to the first AI/ML entrants, but these were mainly based on decision rules programmed for expert systems. The compound annual growth rate (CAGR) of AI-based medical devices over the five-year period from 2015 to 2019 was remarkable, with FDA approvals at 53.6 percent and European CE marks at a 50.4 percent CAGR. By mid-2020, there were over two-hundred AI/ML-based medical devices approved in the United States. The floodgates have opened for medical AI, and healthcare should expect to see AI-augmented decision-making course its way into every facet of medicine in the coming decade. Challenges remain, however, for preventing any harm that might be done by these newly developed algorithms, along with monitoring and improving their performance in diverse clinical settings.

In the United States, the FDA is in charge of regulatory oversight of medical devices. The process for SaMD approval can proceed down three pathways with the FDA. The most rigorous of these is known as the premarket approval pathway (referred to as PMA, designed for high-risk devices). Under the PMA pathway, companies are required to demonstrate reasonable assurance of safety and effectiveness with data typically derived from a controlled trial. The 510(k) pathway is used for lower-risk medical purposes, and it requires that devices demonstrate substantial equivalence to an already-marketed device. The third pathway is de novo premarket review (for low- and moderate-risk devices). The de novo pathway was established to allow new devices to serve as templates, or references, for future 510(k) submissions.

In Europe, regulation is decentralized, and regulatory guidelines are less demanding compared to the FDA. Over the same timeframe, CE marks were granted for 240 AI/ML-based medical device products in Europe. The survey found that 124 of these products were commonly approved in both the United States and Europe. One of the first AI-based algorithms to receive FDA marketing clearance through the 510(k) pathway came for the company Arterys, which built a deep learning algorithm to analyze cardiovascular MRI images in 2016. Arterys followed that up with its Oncology DL product in 2017. In 2018, the FDA cleared IDx's software for diabetic retinopathy under the de novo pathway, one of the first approvals for an AI system that provided screening decisions without the need for clinician interpretation. The IDx product is used to detect severe diabetic retinopathy in adults with diabetes. To date, the AI systems that have been developed and approved are predominantly for radiological imaging applications. The only other specialties with significant representation in AI medical devices are neurology and cardiology, but the application space is on its way to fill in rapidly across the medical spectrum.

Over the short period between introduction of deep learning for medicine until today, the FDA has grappled with how to monitor and regulate AI-based algorithms effectively. Software used in medical devices had previously been required to have "locked" algorithms to ensure that performance, and thus safety standards, were maintained. Under the SaMD umbrella, AI algorithms are considered as clinical decision support software. In a news release from February 2018, the FDA defined AI algorithms as "a type of clinical decision support software that can assist providers in identifying the most appropriate treatment plan for a patient's disease or condition" and emphasized that these algorithms "should not be used as a replacement of a full patient evaluation or solely relied upon to make or confirm a diagnosis." The difficult task facing regulatory agencies is that AI-based algorithms and models present a unique set of challenges and, potentially, a new set of risks for patients.

The adaptive nature of AI algorithms is the key feature that drove the development of a new framework for evaluating SaMDs, and the FDA released guidance documents for discussion in 2019 and an action plan in January 2021.[47,48] A succinct summary is contained in the following excerpt from the framework:

> The traditional paradigm of medical device regulation was not designed for adaptive AI/ML technologies, which have the potential to adapt and optimize device performance in real-time to continuously improve healthcare for patients. The highly iterative, autonomous, and adaptive nature of these tools requires a new, total product lifecycle (TPLC) regulatory approach that facilitates a rapid cycle of product improvement and allows these devices to continually improve while providing effective safeguards.
>
> *US FDA Artificial Intelligence and Machine Learning*
> *Discussion Paper*

The FDA's guidance focused on tracking changes, particularly for architectures, inputs, and the algorithm's potential intended use. Modifications can lead to performance improvement through both the training process and training data. From the FDA's perspective, the question is, what ensures that the new AI system adequately protects patients or maintains safety when put into practice? The recommendation was that companies submit a change control plan and seek premarket approval for changes. In addition, there were also calls for improving transparency (avoiding the "black-box" algorithm problem) and real-world performance monitoring. Much of this squares with the encouragement of best practices in the tech world for AI interpretability, control, debugging, and better metrics.

Overall, the consensus is that the enormous potential and ability of these systems to "learn" in real time will require a regulatory approach that spans the lifecycle of these technologies. There is also an awareness that AI algorithms and models need further improvement for clinical implementation, which will vary by use case. Detecting pneumonia from chest X-rays, one of the most-hyped achievements reported for clinically relevant AI, made use of state-of-the-art technology and 112,000 images. In the study, the deep learning algorithm outperformed a very small set of human experts (AUC 0.76)—four in total. It is questionable how it would fare with a greater number of experts out in the real world, and the accuracy is so low that it arguably would not pass muster by regulators. A final point is that the AI engineering and medicine community would benefit from standards around ground truth datasets and performance testing. Challenges such as Camelyon (AI for metastatic breast cancer[49]) and others have advanced development, assembled datasets similar to ImageNet, and put forward benchmarks.

Deep learning's impact on radiology and clinical oncology are dramatic, enabling intelligent machine-based tumor detection, subtyping and grading, while advancing novel tasks such as mutation detection, survival, and response prediction from a wide range of data types alone or in combination. Over the next decade, AI-based algorithms and medical practice will overlap and eventually converge to transform medicine. Before that vision can be realized, a number of challenges, summarized in the following feature, will need to be addressed.

CHALLENGES FOR THE IMPLEMENTATION PHASE OF CLINICAL AI

Generalization: An important step for any AI trained on hundreds of thousands of examples from baseline or canonical datasets is the demonstration, or proof, of the underlying algorithm's ability to generalize. The requirement means that an AI tool must perform well when applied under conditions on which it was not trained, such as with images obtained on new devices or patient populations. In other words, the AI system needs to operate with some flexibility in the face of technical and biological variation. The algorithm should not suffer any significant loss of performance and also not require extensive retraining. Poor generalizability is often the result of model overfitting on the training data.

Validation: The current phase of AI research in medicine must shift toward building evidence around clinical efficacy in prospective trials and further replication in independent sample sets. A challenge in cancer research is the dependence of training AI models on TCGA data sets that are limited to academic researchers. A similar concern holds true for neurology and its academic image repositories. A second challenge for validation is the need for rigorous testing of the frameworks and algorithms in real-world settings employing randomized, controlled trials to overcome adoption barriers and FDA/CE regulatory requirements.

Applicability: AI built in research laboratories may not be suited for real-world problems or designed for clinical evaluation and deployment. The AI systems make predictions, but often there is no measure of confidence reported, which is needed for evaluation. Scientists will need to work with physicians and laboratory directors to understand clinical workflows to implement solutions that benefit decision-making in practice.

Explainability: There is a desire within the medical community and also at the regulatory agency level to address the "black-box" problem. To make diagnostic decisions or therapy recommendations, doctors may require AI systems to produce an underlying feature in the data as to why classifications or predictions were made. Hospitals may be responsible for decisions an AI makes, so interpretability will become mandatory. Explainability is now an implicit requirement by EU member states under the General Data Protection Regulation (GDPR), along with a right to understand.

Causation: Nearly all machine learning approaches employ associative inference for diagnosis or when assigning probabilities for therapeutic decisions; they are not able to identify cause and effect. Medical practice relies on doctors using causal reasoning to identify the best explanations for a set of symptoms or therapeutic effect.

> **Bias and fair access:** Algorithms that process personal data for use in decision-making by health systems face a challenge in representing the population. AI algorithms built for skin cancer do not perform equally given inputs with varying skin tone, as one example. Clinical trials that employ AI will likely be trained on racially or culturally biased data. This consideration needs to be addressed at the planning stages of AI systems development. A related concern is equitable access to modern AI tools for medicine. Science and society will need to decide how to widen access to AI-enhanced healthcare.

Notes

1. Topol EJ. High-performance medicine: the convergence of human and artificial intelligence. Nat Med. 2019 Jan;25(1):44–56.

2. Longoni C, Bonezzi A, Morewedge CK. Resistance to Medical Artificial Intelligence. J Consum Res. 2019 Dec 1;46(4):629–50.

3. Muse ED, Godino JG, Netting JF, Alexander JF, Moran HJ, Topol EJ. From second to hundredth opinion in medicine: A global consultation platform for physicians. NPJ Digit Med. 2018 Oct 9;1:55.

4. Segal MM, Abdellateef M, El-Hattab AW, Hilbush BS, De La Vega FM, Tromp G, et al. Clinical Pertinence Metric Enables Hypothesis-Independent Genome-Phenome Analysis for Neurologic Diagnosis. J Child Neurol. 2015 Jun;30(7):881–8.

5. Richens JG, Lee CM, Johri S. Improving the accuracy of medical diagnosis with causal machine learning. Nat Commun. 2020 Aug 11;11(1):3923.

6. Pierson E, Cutler DM, Leskovec J, Mullainathan S, Obermeyer Z. An algorithmic approach to reducing unexplained pain disparities in underserved populations. Nat Med. 2021 Jan;27(1):136–40.

7. Kim DH, MacKinnon T. Artificial intelligence in fracture detection: transfer learning from deep convolutional neural networks. Clin Radiol. 2018 May;73(5):439–45.

8. Yala A, Mikhael PG, Strand F, Lin G, Smith K, Wan Y-L, et al. Toward robust mammography-based models for breast cancer risk. Sci Transl Med. 2021 Jan 27;13(578).

9. Titano JJ, Badgeley M, Schefflein J, Pain M, Su A, Cai M, et al. Automated deep-neural-network surveillance of cranial images for acute neurologic events. Nat Med. 2018 Sep;24(9):1337–41.

10. Mobadersany P, Yousefi S, Amgad M, Gutman DA, Barnholtz-Sloan JS, Vega JEV, et al. Predicting cancer outcomes from histology and genomics using convolutional networks. Proc Natl Acad Sci. 2018 Mar 27;115(13):E2970–9.

11. Eitel F, Soehler E, Bellmann-Strobl J, Brandt AU, Ruprecht K, Giess RM, et al. Uncovering convolutional neural network decisions for diagnosing multiple sclerosis on conventional MRI using layer-wise relevance propagation. NeuroImage Clin. 2019;24:102003.

12. Zhao Y, Wang T, Bove R, Cree B, Henry R, Lokhande H, et al. Ensemble learning predicts multiple sclerosis disease course in the SUMMIT study. Npj Digit Med. 2020 Oct 16;3(1):1–8.

13. Poon CCY, Jiang Y, Zhang R, Lo WWY, Cheung MSH, Yu R, et al. AI-doscopist: a real-time deep-learning-based algorithm for localising polyps in colonoscopy videos with edge computing devices. Npj Digit Med. 2020 Dec;3(1):73.

14. Kiani A, Uyumazturk B, Rajpurkar P, Wang A, Gao R, Jones E, et al. Impact of a deep learning assistant on the histopathologic classification of liver cancer. NPJ Digit Med. 2020 Feb 26;3:1-8.

15. Li D, Bledsoe JR, Zeng Y, Liu W, Hu Y, Bi K, et al. A deep learning diagnostic platform for diffuse large B-cell lymphoma with high accuracy across multiple hospitals. Nat Commun. 2020 Dec;11(1):6004.

16. Bulten W, Pinckaers H, van Boven H, Vink R, de Bel T, van Ginneken B, et al. Automated deep-learning system for Gleason grading of prostate cancer using biopsies: a diagnostic study. Lancet Oncol. 2020 Feb 1;21(2):233–41.

17. Ström P, Kartasalo K, Olsson H, Solorzano L, Delahunt B, Berney DM, et al. Artificial intelligence for diagnosis and grading of prostate cancer in biopsies: a population-based, diagnostic study. Lancet Oncol. 2020 Feb 1;21(2):222–32.

18. Cornblath EJ, Robinson JL, Irwin DJ, Lee EB, Lee VM-Y, Trojanowski JQ, et al. Defining and predicting transdiagnostic categories of neurodegenerative disease. Nat Biomed Eng. 2020 Aug;4(8): 787–800.

19. Narayanan H, Dingfelder F, Butté A, Lorenzen N, Sokolov M, Arosio P. Machine Learning for Biologics: Opportunities for Protein Engineering, Developability, and Formulation. Trends Pharmacol Sci. 2021 Mar;42(3):151–65.

20. Hannun AY, Rajpurkar P, Haghpanahi M, Tison GH, Bourn C, Turakhia MP, et al. Cardiologist-Level Arrhythmia Detection and Classification in Ambulatory Electrocardiograms Using a Deep Neural Network. Nat Med. 2019 Jan;25(1):65–9.

21. Landi I, Glicksberg BS, Lee H-C, Cherng S, Landi G, Danieletto M, et al. Deep representation learning of electronic health records to unlock patient stratification at scale. Npj Digit Med. 2020 Dec;3(1):96.

22. Hofer IS, Lee C, Gabel E, Baldi P, Cannesson M. Development and validation of a deep neural network model to predict postoperative mortality, acute kidney injury, and reintubation using a single feature set. Npj Digit Med. 2020 Dec;3(1):58.

23. Capper D, Jones, DTW, Sill M, Hovestadt V, Schrimpf D. DNA methylation-based classification of central nervous system tumours. Nature. 2018 Mar 22;555(7697):469–74.

24. Jurmeister P, Bockmayr M, Seegerer P, Bockmayr T, Treue D, Montavon G, et al. Machine learning analysis of DNA methylation profiles distinguishes primary lung squamous cell carcinomas from head and neck metastases. Sci Transl Med. 2019 Sep 11;11(509).

25. Jiao W, Atwal G, Polak P, Karlic R, Cuppen E, Danyi A, et al. A deep learning system accurately classifies primary and metastatic cancers using passenger mutation patterns. Nat Commun. 2020 Feb 5;11(1):728.

26. Huang S-C, Pareek A, Zamanian R, Banerjee I, Lungren MP. Multimodal fusion with deep neural networks for leveraging CT imaging and electronic health record: a case-study in pulmonary embolism detection. Sci Rep. 2020 Dec;10(1):22147.

27. Trebeschi S, Drago SG, Birkbak NJ, Kurilova I, Călin AM, Delli Pizzi A, et al. Predicting response to cancer immunotherapy using noninvasive radiomic biomarkers. Ann Oncol Off J Eur Soc Med Oncol. 2019 Jun 1;30(6):998–1004.

28. Kather JN, Heij LR, Grabsch HI, Loeffler C, Echle A, Muti HS, et al. Pan-cancer image-based detection of clinically actionable genetic alterations. Nat Cancer. 2020 Aug;1(8):789–99.

29. Fu Y, Jung AW, Torne RV, Gonzalez S, Vöhringer H, Shmatko A, et al. Pan-cancer computational histopathology reveals mutations, tumor composition and prognosis. Nat Cancer. 2020 Aug;1(8): 800–10.

30. Hoadley KA, Yau C, Hinoue T, Wolf DM, Lazar AJ, Drill E, et al. Cell-of-Origin Patterns Dominate the Molecular Classification of 10,000 Tumors from 33 Types of Cancer. Cell. 2018 Apr;173(2):291-304.e6.

31. Gensheimer MF, Aggarwal S, Benson KRK, Carter JN, Henry AS, Wood DJ, et al. Automated model versus treating physician for predicting survival time of patients with metastatic cancer. J Am Med Inform Assoc. 2020 Dec 14.

32. Golia Pernicka JS, Gagniere J, Chakraborty J, Yamashita R, Nardo L, Creasy JM, et al. Radiomics-based prediction of microsatellite instability in colorectal cancer at initial computed tomography evaluation. Abdom Radiol. 2019 Nov 1;44(11):3755–63.

33. Kather JN, Pearson AT, Halama N, Jäger D, Krause J, Loosen SH, et al. Deep learning can predict microsatellite instability directly from histology in gastrointestinal cancer. Nat Med. 2019 Jul;25(7):1054–6.

34. Schaumberg AJ, Rubin MA, Fuchs TJ. H&E-stained Whole Slide Image Deep Learning Predicts SPOP Mutation State in Prostate Cancer. Pathology; 2016 Jul.

35. Coudray N, Ocampo PS, Sakellaropoulos T, Narula N, Snuderl M, Fenyö D, et al. Classification and mutation prediction from non–small cell lung cancer histopathology images using deep learning. Nat Med. 2018 Oct;24(10):1559–67.

36. Szegedy C, Vanhoucke V, Ioffe S, Shlens J, Wojna Z. Rethinking the Inception Architecture for Computer Vision. ArXiv151200567 Cs. 2015 Dec 11.

37. Zheng H, Momeni A, Cedoz P-L, Vogel H, Gevaert O. Whole slide images reflect DNA methylation patterns of human tumors. Npj Genomic Med. 2020 Mar 10;5(1):1–10.

38. Szegedy C, Joffe S, Vanhoucke V, Alemi A. Inception-v4, Inception-ResNet and the impact of residual connections on learning. Proc Thirty-First AAAI Conf Artif Intell AAAI Press. 2017; 4:4278–84.

39. Wen PY, Watson D, Kapoor S, Alam A, Alam A, Lala DA, et al. Superior therapy response predictions for patients with glioblastoma (GBM) using Cellworks Singula: MyCare-009-03. J Clin Oncol. 2020 May 20;38(15_suppl):2519–2519.

40. Image and Data Archive, University of Southern California. Available from: `ida.loni.usc.edu/login.jsp?project=ADNI`.

41. Wan J, Zhang Z, Yan J, Li T, Rao BD, Fang S, et al. Sparse Bayesian multi-task learning for predicting cognitive outcomes from neuroimaging measures in Alzheimer's disease. In: 2012 IEEE Conference on Computer Vision and Pattern Recognition. 2012. p. 940–7.

42. Davatzikos C, Fan Y, Wu X, Shen D, Resnick SM. Detection of prodromal Alzheimer's disease via pattern classification of magnetic resonance imaging. Neurobiol Aging. 2008 Apr 1;29(4):514–23.

43. Stonnington CM, Chu C, Klöppel S, Jack CR, Ashburner J, Frackowiak RSJ. Predicting clinical scores from magnetic resonance scans in Alzheimer's disease. NeuroImage. 2010 Jul 15;51(4):1405–13.

44. Ding Y, Sohn JH, Kawczynski MG, Trivedi H, Harnish R, Jenkins NW, et al. A Deep Learning Model to Predict a Diagnosis of Alzheimer Disease by Using 18F-FDG PET of the Brain. Radiology. 2018 Nov 6;290(2):456–64.

45. Bakshi R, Healy BC, Dupuy SL, Kirkish G, Khalid F, Gundel T, et al. Brain MRI Predicts Worsening Multiple Sclerosis Disability over 5 Years in the SUMMIT Study. J Neuroimaging. 2020;30(2):212–8.

46. Muehlematter UJ, Daniore P, Vokinger KN. Approval of artificial intelligence and machine learning-based medical devices in the USA and Europe (2015–20): a comparative analysis. Lancet Digit Health. 2021 Mar 1;3(3):e195–203.

47. US FDA. Proposed regulatory framework for modifications to artificial intelligence/Machine Learning [AL/ML]-based software as a medical device [SaMD]. Available from: `www.fda.gov/media/122535/download`.

48. US FDA. Artificial Intelligence/Machine Learning (AI/ML)-Based Software as a Medical Device {SaMD) Action Plan. 2021 Jan. Available from: `www.fda.gov/media/145022/download`.

49. Ehteshami Bejnordi B, Veta M, Johannes van Diest P, van Ginneken B, Karssemeijer N, Litjens G, et al. Diagnostic Assessment of Deep Learning Algorithms for Detection of Lymph Node Metastases in Women With Breast Cancer. JAMA. 2017 Dec 12;318(22):2199.

AI in Drug Discovery and Development

Under normal conditions the research scientist is not an innovator but a solver of puzzles, and the puzzles upon which he concentrates are just those which he believes can be both stated and solved within the existing scientific tradition.

—Thomas S. Kuhn, *The Structure of Scientific Revolutions*

The desire for *in silico* methods to simulate biological phenomena or to model the pharmacological properties of small molecule drugs has existed since the dawn of the computer age. Although the phrase *in silico biology* was probably first used in the late 1980s, it is apparent from the writings of the early AI pioneers that neuron-like, or even brain-level, simulations were contemplated with the simple computational devices that became available in the early 1950s. In pharmacology and drug discovery, computational methods for deriving structure-activity relationships also date back to the 1950s with the foundational work of Corwin Hansch. The now ubiquitous use of *quantitative structure-activity relationship (QSAR)* models stems from Hansch's equations. Efforts to use computation for supporting organic chemical synthesis started in the 1960s.[1] Once computing languages and personal computers became available for nonexpert programmers, software programs were written to study chemical reactions and biological phenomena. One of the first for *in silico* biology was written to model blood cell metabolic pathways on a Mac computer.[2] More recently, sophisticated simulations of the entirety of bacterial and eukaryotic cell function have been created.

It has long been a dream to do biology and chemistry entirely *in silico*. Organic chemistry will likely cross the *in silico* finish line first, long before machine-based predictions of cellular or systems-level biological phenomena are established. The rules and grammar of chemistry, plus Schrödinger's equations, align in computational chemistry's favor. The latest *in silico* tools for enumerating compound libraries and predicting synthesis routes will work together to create a vast catalog of nearly any synthesizable small organic compound out of the immensity of *chemical space*. Coupling these technologies with AI systems for searching the chemical space to pinpoint ideal clinical compound candidates, based on target 3D structure and predicted drug-target binding affinities, may turn the chemical laboratory into a highly efficient drug engineering factory. The missing component is target discovery and biological hypotheses around targets and their causal roles in disease or pathophysiological mechanisms. Pharma as an industry has largely left the details of understanding disease biology to academic researchers. However, identifying targets and understanding the biological context of any drug action are paramount to clinical success. The hope is that biology can become more like chemistry, once a fairly complete corpus of knowledge is in hand. This would enable AI systems to be used routinely to decipher complex biology and run *in silico* experiments with high predictive accuracy.

For the drug hunters, the current and future applications of AI are an Alice in Wonderland scenario. The question "Which way should we go?" does indeed depend on where we are headed. A goal of end-to-end autonomous drug design, where AI also predicts the safety and efficacy of a drug with a high degree of certainty before any clinical trials are launched, is laudable but entirely out of reach. No one is suggesting this today, but that day will come. Across big pharma, more strategic efforts are focusing on validation of AI technology and narrow wins, such as implementing an AI tool that augments small molecule hit or lead decision-making, as with diagnostic or clinical decision support applications. More ambitious goals for reproducible AI systems do envision autonomous AI-guided decisions around hit selection. The business rationale is that AI will improve hit rates, select or rank-order the best targets from data, and shrink the timeline from target discovery to a pre-clinical candidate. As with the introduction of any new technology, this will be a slow process with a watchful eye on whether AI can improve overall productivity in pharma.

Success with AI in drug discovery will require some rethinking of the traditional approaches toward tackling problems and retooling of processes for developing drugs. AI today can serve as a powerful hypothesis engine, especially for synthetic chemistry. Can chemists shift from current tools and intuition to build desired compounds after consulting an AI oracle? Across early discovery R&D, scientists will need to pay attention to how data should be generated and formatted for AI model consumption. AI integration into drug discovery has fewer hurdles than the use of AI-powered algorithms for medical imaging and diagnostics applications. No harm is done by a poor decision in the early discovery phases in an R&D laboratory except to the pharmaceutical company's efficiency and bottom line, as there are ample checks and balances along the development path. On the flip side, there are many who anticipate productivity gains, should AI prove as successful in pharma as it has in other industries. Industry executive Mark Murcko of Dewpoint Therapeutics puts it this way: "Imagine what that world will look like five years from now, when the libraries are trillions of compounds, the scoring functions are quite good and the cost of synthesis is quite low. This starts to feel fairly powerful."[3]

A Brief Survey *of In Silico* Methods in Drug Discovery

The great advances in biochemistry and molecular biology, the development of physical organic chemistry, and the availability of large computers are creating opportunities for restructuring medicinal chemistry. The enormous volume of scientific results relevant to medicinal chemistry which appear with each new round of the journals forces us to make greater efforts to bring the information together in more meaningful patterns. The QSAR paradigm redirects our thinking about structuring medicinal chemistry.

— Corwin Hansch, 1976

Pharmacological modeling and *in silico* drug discovery have advanced primarily from incorporating modern computing and statistical techniques to exploit newly available data-generating technologies and the exponential growth of compound databases. Cheminformatics techniques have attained far better accuracy and utility over successive generations, much like other scientific disciplines that have adopted analytical and knowledge-driven approaches. The evolution of *in silico* approaches, and in particular computer-assisted drug design, has occurred through the development of an array of methods that include QSAR modeling in all of its forms (2D, 3D, and others). Chemists use QSAR models to predict the biological activities of chemical structures. For the former, this is done by analysis of quantitative characteristics of structural features, where a mathematical model is constructed to relate a molecular structure to a pharmacological property using statistical techniques. QSAR derivation and validation has a checkered history because of some of the known weaknesses of machine learning algorithms on any type of data, the main problems being overfitting and overly broad application. QSAR builders circa the 1990s had fallen into the trap of applying the outputs of these models as hard filters for datasets that were not covered in the training data, a deficiency of the chemical databases available at the time.

Early QSAR models used basic multivariate regression models to correlate potency (for inhibitors, typically given as log IC_{50}) with substructure motifs and chemical properties like molecular weight, solubility (log P), hydrophobicity, and other physiochemical factors. QSARs use molecular descriptors as numerical representations of chemical structures. There are many molecular descriptors, and these get classified according to the dimensionality of the chemical representation from which they are computed. For example, descriptors encoding numerically generic chemical properties provide straightforward estimations of the size, structure, and lipophilicity of molecules. Despite their low dimensionality, some of these descriptors are tightly associated with predicting drug-like molecules enshrined in Lipinski's rule of five. The rules form guidelines for candidate compound selection and library pre-screening. Molecules conforming to three out of four of the following satisfy Lipinski's criteria: molecular mass under 500 Daltons; lipophilicity where Log $P < 5$; hydrogen bond donors, less than 5; hydrogen bond acceptors, less than 10.[4] Drug developers also include molar refractivity, a measure of polarity in which a molecule should fall within the range of 40 to 130.

Theoretically, a QSAR model can predict the target property for any compound for which chemical descriptors can be calculated. However, if a compound is highly dissimilar from all other compounds used to train the model, reliable prediction of its activity is unlikely. Model performance can be improved by going beyond simple 1D molecular descriptors. The use of 2D descriptors broadens the chemical space of drug-like compounds, as was found after analysis of a collection of anti-cancer drugs.[5] A final comment is that the linearity assumptions made in QSARs impose limitations on model performance, a problem that can be surmounted by deep learning. Despite these known limitations, QSARs have been successful and a great improvement over the earlier guesswork involved in linking chemical properties to biological actions.

Beyond QSAR, the cheminformatics toolkit has vastly expanded. Software is widely available for binding site prediction, molecular docking, virtual screening, ligand design, similarity searching, binding free energy estimation, ADMET prediction, and pharmacophore, as well as molecular dynamics modeling and extensive use of statistics-based data analysis tools. Since the 1960s, researchers have also endeavored to develop computational methods for computer-aided drug synthesis. It is anticipated that prediction of synthetic routes and use of generative adversarial networks will push the boundaries of what is achievable with AI in drug design.

Cheminformatics tools are now critical at the starting point for drug design and screening projects. Robust software now exists for computationally generating virtual compound libraries by various enumeration techniques to greatly expand chemical space exploration. One of the first was the creation of GDB-17, a small molecule library with 160 billion organic compounds containing up to 17 atoms. Big pharma has produced extraordinarily sized libraries, including Lilly's Proximal Collection (10^{10} compounds), Merck's Accessible inventory MASSIV (10^{20}), and AstraZeneca's library based on Enamine's database (10^{17}), to name a few. Software such as KNIME and DataWarrior (both Open Source), MolSoft, or Reactor (ChemAxon) are accessible tools to enumerate chemical libraries.

Drug design and *in silico* screening paradigms have improved in lock step with computing power and the exponential growth in structural databases for small molecules and proteins. One of importance to drug makers has been the Cambridge Structural Database, a repository of small molecule structures derived from crystallography. These structures provide critical information on bond lengths and angles, in addition to torsion angles on which force-fields are built. The data provide quantitative insights on potential intermolecular interactions needed to design high affinity drugs, such as hydrogen bonding. The database has accumulated more than one million structures from X-ray and neutron diffraction analyses. A second essential resource is the Protein Data Bank (PDB), a database that stores more than 150,000 experimentally obtained 3D structures of biological macromolecules. The structures were determined predominantly by X-ray diffraction, solution NMR, and electron microscopy methodologies. *Structure-based virtual screening (SBVS)* programs and drug design use 3D structures from PDB and other sources in small molecule drug discovery. A subset of PDB, PDBbind, contains a curated set often used for biophysics and machine learning techniques for molecular docking. For QSAR studies, pharmaceutical companies leverage internal or external resources that contain log Ps and pK_as for candidate compounds, plus a database with tens of thousands of Hammett equations for physical organic reactions and Hansch equations for bioactivities. Beyond these common resources, a formidable collection of information on other databases is found on the Swiss Institute of Bioinformatics' website, `click2drug.org`. In addition, the collection includes a comprehensive directory of other *in silico* drug design tools.

Pharmaceutical R&D employs cheminformatics tools widely, where many are used to drive the design-make-test-analyze (DMTA) cycle that is prevalent across the industry. This time-consuming and iterative process consists of designing new compounds, performing chemical synthesis, testing selected compounds in biological assays, and analyzing the results before starting the next round of designs. In the design phase (lead optimization), chemists use QSAR or qualitative SAR models to make predictions around the biological activities of chemical structures.

Virtual Screening with Cheminformatics and HTS Technologies

Modern pharmaceutical laboratories operate *high-throughput screening (HTS)* platforms to evaluate up to a few million compounds in miniaturized assay formats, a process that is largely driven by robotic automation. Physical compound screening with HTS can be carried out using well-engineered assays that identify candidates via biochemical interactions, cellular morphology or phenotype, or model organism-based outputs. Laboratory HTS is often the beginning of a drug discovery and development program. The throughput ceiling (10^6 to 10^7 compounds) and the restriction it imposes on interrogating the potential chemical space of drug-like molecules has made *in silico* virtual screening an essential tool in pharma. Although the progress in DNA-encoded library technology potentially allows far larger physical libraries to be interrogated ($10^9 - 10^{12}$), it requires expertise and is limited to aqueous synthesis conditions. Regardless, *in silico* target or ligand-based virtual screening is now an embedded and highly efficient process in drug discovery.

In contrast to technology-driven HTS, virtual screening is a knowledge-driven approach, requiring structural information on the drug target (SBVS) or on bioactive ligands for the target, referred to as *ligand-based virtual screening*.[6] *In silico* library screening emerged in the 1980s. After a decade of development, it became a more widely used tool that predictably followed the increase in computational power and growth in structural data of target molecules. Virtual screens can appear at multiple stages of the discovery and development pipeline, but often are the starting points to score and rank billions of molecules found in large chemical libraries or virtual databases. Many drugs, including Captopril (antihypertensive), Tirofiban (antiplatelet), Indinavir and Ritonavir (HIV), and Dorolamide (glaucoma), were discovered by virtual screening.

Figure 7.1 shows an example of SBVS combined with downstream HTS to obtain lead compounds. The starting point is either a 3D structure of the target protein or a proxy structure derived by performing homology modeling for a target protein lacking a crystal structure. Virtual databases or compound libraries are pre-filtered to obtain subsets with drug-like properties or to eliminate difficult structures or those with known toxicities. Molecular docking software can score candidate compounds based on how well its structure fits into the binding pocket of the respective target protein. Further computational scoring and analysis techniques such as hierarchical clustering, principal components analysis, and other ML methods can analyze and break down large molecular libraries into smaller collections of similar compounds by rank. A final step identifies the top-ranked molecules as hits or leads. The largest published structure-based virtual screening experiment reported docking of 170 million compounds, identifying 453,000 ligands for the dopamine D4 receptor, 30 of which bound the target with high affinity.[7]

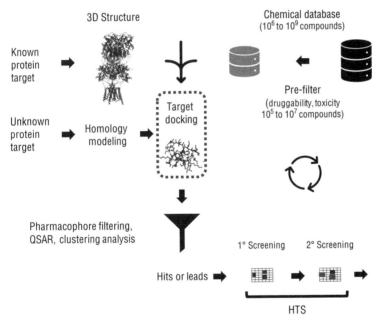

Figure 7.1: A structure-based virtual screening workflow

One of many possible workflows is shown for structure-based virtual screening. Virtual screens vastly improve the search of chemical space while decreasing the time to arrive at a clinical candidate.

Primary HTS assays placed downstream of virtual screening are designed to measure the binding of a compound to the protein target, detect compound agonist or antagonist activity, or measure alterations to a cellular phenotype.

Secondary screens can be designed to detect potential toxicity. At this juncture, significant medicinal chemistry is still required to optimize compounds into leads by improving selectivity, *in vivo* exposure, and further reduce toxicity. Often, this is the beginning of another DMTA cycle, which might require testing of hundreds to thousands of molecules to arrive at a clinical candidate with optimal biological activity and low toxicity profile.

Drug development activities take many routes to a new product, and the ability to discover drug-like molecules that function similarly to others is a common path. A ligand-based approach is taken, for example, to circumvent undesirable properties that cause adverse events in known drugs. Alternatively, a program might be initiated to discover compounds that mimic the favorable properties of established compounds. Computer-assisted drug design can be employed to suggest candidates that can then be experimentally validated. The development of ligand-based virtual screening followed a similar trajectory as other *in silico* approaches, performed until recently with standard statistical methods. Chemical similarity calculations form the core of ligand-based screens. Ligand virtual screens, however, are very amenable to machine learning methodology, since known ligands can form the basis of a training set to discriminate similar from dissimilar compounds with a predictive model. Similarity searches based on compound "likeness" alone have not fared as well as those that also incorporate structure information of the target. Scoring and ranking of ligand screening results are typically much better but at the cost of increased prediction time, and the *in silico* process is suitable only when a 3D structure of the protein and binding pocket are available. Several types of ligand-based screening methods exist, based on the type and availability of structural information being used.

AI Brings a New Toolset for Computational Drug Design

For the past 60 years, experts have been trying to dictate the rules of chemistry to computers via hand-coded heuristics. Instead, we anticipate that equipping machines with strong, general planning algorithms, symbolic representations, and the means to learn autonomously from the rich history of chemistry will be crucial to allowing the machine to become accepted as a valuable assistant in chemical synthesis.

— Marwin Segler, Mark Preuss, and Mark Waller, *Nature* 2018

Successful integration of AI tools and associated research methodologies into pharmaceutical R&D will depend on demonstrable proof that AI can either readily outperform built-in conventional capabilities or do new tricks to increase scale, efficiency, and interpretability. Drug design necessitates both *forward synthesis*, the process of taking chemical building blocks and predicting reaction conditions to create new molecules, and *retrosynthesis*, which works backwards from the final desired product, as its name implies. In retrosynthesis, the task is to find synthetic paths that lead to a desired compound by creating the tree of possible alternative routes, using building blocks and intermediate reaction steps that fall within the rules of chemistry.

Currently, medicinal chemists rely on instinct and biases that favor known and robust reactions to achieve chemical transformation by retrosynthetic routes. These boil down to a limited set of reactions that are broadly applied to structurally diverse substrates, can be completed relatively quickly, and can be conducted on laboratory-friendly, standard equipment—essentially, a chemist's version of faster, better, cheaper. What may come as a surprise to those outside of the small molecule-driven Pharma universe is that just five types of organic chemical reactions account for greater than 60 percent of all of those ever used to produce compounds for drug discovery.[8] The workhorse reactions that dominate are amide formation, the Suzuki–Miyaura reaction, aromatic nucleophilic substitutions, amine Boc-deprotection, and electrophilic reactions with amines. To put this into perspective, there are likely hundreds of millions of synthetic routes to produce the estimated 10^{60} synthetically feasible, drug-like, small organic molecules (with molecular weight below 500 Daltons). It is impossible to reach even a small fraction of a percent of this chemical space with five reaction types, but these form the basis of large-scale manufacturing activity for small molecule drugs in the pharmaceutical industry.

Reviews of pharmaceutical patent literature describe more than one million reaction types. The astronomical number of potential small molecule drugs and the staggering number of available chemical synthesis routes to build them seem completely at odds with the discovery process in the pharma industry. Although it can be argued that strict chemical and structural limitations (plus practical considerations in chemical manufacturing) are in place to vastly reduce the small molecule universe, the fact that the small molecule drug "discovery" paradigm seems almost deliberately set up *not* to explore new synthesis routes is a bewildering paradox.

The conventional approach thus has obvious drawbacks in attempting to identify optimal, high-quality drug candidates or for discovering leads to challenging targets. Furthermore, the use of compound libraries that are filtered to deliver initial hit structures imposes constraints on the type of chemistry pursued downstream. The focus on ease of synthesis has roots not only in the underlying chemistry process but also in the research culture and necessities of hitting company milestones and personal performance metrics.

One of the clearest examples of how AI systems could be built to perform tasks previously unattainable by other computational methods is found in recent work by Marwin Segler and colleagues on organic chemical retrosynthesis. The ability to construct viable synthetic pathways with known chemical precursors is in some ways analogous to the task of medical diagnosis. Arriving at a solution requires many steps and differential decisions. In chemistry, the combinatorics greatly favor an intelligent machine. Instead of AI threatening the job of a trained physician, it might well be the case that a well-trained AI system might replace both synthetic and medicinal chemists one day. The development of a modular AI system comprised of multiple neural networks combined with symbolic AI logic shows a powerful way forward.

Computational approaches toward retrosynthesis use rules-based design tools and application of machine learning have seen limited use. Three software packages are commercially available for retrosynthetic planning using expert-based rules, those being IC*SYNTH* (DeepMatter), Synthia (from Merck KGaA; formerly Chematica), and ChemPlanner (Wiley). The modular process proposed by Segler and colleagues takes advantage of three neural networks combined with Monte Carlo tree search, which was named 3N-MCTS.[9] The use of MCTS for the retrosynthesis problem is a natural fit, since the process constructs a decision tree of probabilities and weights, determining possible completion sequences and favorable outcomes. The innovative step is that MCTS's completion behavior is learned as a neural network. Other groups have investigated this approach for generative chemistry. The network is trained on the entire known corpus of organic chemistry reactions. Rules were automatically extracted from 12.4 million reactions out of the Reaxys chemistry database, resulting in a rule set of approximately 300,000 that were used in two different phases of the pipeline. In tests of retrosynthesis capabilities, the 3N-MCTS method outperformed state-of-the-art search algorithms in both structure-solving tests and time to completion. The deep learning technology was acquired by Elsevier, creator of the Reaxys database, and now offered as its Reaxys Predictive Retrosynthesis product.

Accurate prediction of organic molecule synthesis routes has been achieved by several machine learning techniques in the past decade. More recently, these include the use of logic and knowledge graphs,[10] neural networks,[11] and machine translation.[12,13] One of the keys to further development of AI approaches such as 3N-MCTS will be to discover and curate more reaction data, which can be found in abundance but is strewn across electronic laboratory notebooks and patents from chemical, pharmaceutical, and academic research settings.

A second game-changing approach coming out of the AI research community that may have staying power for synthesis prediction tasks in chemistry is the use of *natural language processing (NLP)* models. Figure 7.2 shows a schematic that draws an analogy between language translation and chemistry.

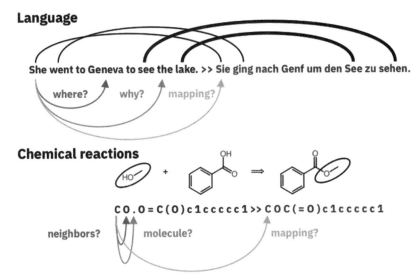

Figure 7.2: Analogous mapping in language translations and chemical reactions

Top: A mapping between an English phrase and the German translation. Bottom: A mapping between reactants (methanol + benzoic acid) and a product molecule (methyl benzoate) in a chemical reaction represented with a text-based line notation called SMILES.

Source: Used with permission from Philippe Schwaller, IBM Research (CC BY-ND 2.0), https://www.ibm.com/blogs/research/2021/04/rxnmapper-chemistry-grammar/.

The pace of advancement in NLP tools in science has been astonishing, pouring over into chemistry, biology, and physics in rapid succession. The use of architectures such as the Transformer model for language understanding applied to chemistry's grammar has been pioneered by Phillipe Schwaller and colleagues at IBM Research and the University of Bern, Switzerland.[14] The task directed at the unsupervised training method is to learn how atoms rearrange during reactions.* The atom-mapping task is central to computational chemistry and in the past required human input and manual curation. Tools such as DREAM, AutoMapper, and Indigo are examples. The tool that was built by the team, RXNMapper, can perform atom-mapping on millions of reactions in a few hours on large datasets and build chemical reaction rules for synthesis. The great advantage of this approach is its ability to learn the atom-mapping signal in an unsupervised manner. Figure 7.3 shows an overview of the NLP Transformer-based system behind RXNMapper.

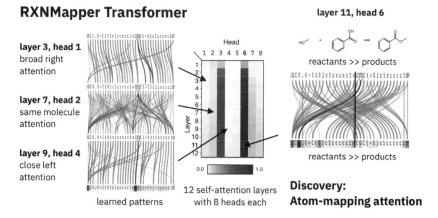

Figure 7.3: Overview of the RXNMapper Transformer tool

The RXNMapper uses Transformer-based architecture to model chemical grammar, as opposed to its typical use in language understanding tasks. Self-attention modules are used in many of the BERT Transformers. An attention function can be described as mapping a query and a set of key-value pairs to an output, where the query, keys, values, and output are all vectors. Here the system builds representations based on the context of atoms and attempts to connect the concepts together. The molecular representation is in simplified molecular-input line-entry system (SMILES), a specification for describing the structure of chemical species using short ASCII strings. Examples of learned patterns (left), the matrix of layers X heads (middle), and an atom mapping of reactants to products (right) are shown.

Source: Figure used with permission and courtesy of Philippe Schwaller, IBM Research (CC BY-ND 2.0), https://www.ibm.com/blogs/research/2021/04/rxnmapper-chemistry-grammar/.

RXNMapper displayed star quality performance, recording a 99.4 percent accuracy on nearly 50,000 atom mapping tasks and also outperforming best-in-class atom mapping algorithms. Of importance to industry, the underlying method is highly scalable, without the need for labeled training data or an engineered set of expert rules.

---

*Atom mapping is the process of deriving a one-to-one mapping between the reactant and product atoms in a chemical reaction.

AI-Based Virtual Screening Tools

The challenging computational task of predicting binding affinities for target-ligand complexes has led to research into applying deep neural network–based approaches as tools for SBVS. At present, there are three approaches that appear promising as an alternative to computationally expensive biophysics-based scoring methods. These methods rely on different types of CNNs, or their use in combination. The 3D-CNN models use a voxelized representation of atoms in 3D space. Spatial graph CNNs exploit graph representations of protein-drug compound complexes. Both styles portray compounds in significantly different ways and use distinct mechanisms of inference for obtaining predictions of binding free energy. A third method, deep fusion, combines features from 3D-CNNs and spatial graph CNNs to improve predictions of binding affinity.

A deep fusion model put into practice during the COVID-19 pandemic illustrates the power of the combination and continued need for model development. A research group formed across several of the US Department of Energy's national laboratories, plus NVIDIA, published results from an SBVS after testing more than 500 million small molecules against four protein structures from SARS-CoV-2 virus. The massive screening effort evaluated more than 5 billion docked poses on these SARS-CoV-2 targets.[15] This in-depth study presented performance metrics comparing computational predictions from two scoring methods to its Fusion model from compounds that were selected and tested experimentally. For two of the four targets, the Coherent Fusion method performed best, but the other methods ranked higher in correlating with experimental results for the other two. These data underscored the difficulty in building CNN-based scoring systems compared to previously developed, biophysics-based methods. The main advantage with the Coherent Fusion model was the throughput resulting from the speed increase that can be achieved, which was 2.7x versus Vina and 403x versus MM/GBSA methods. The Coherent Fusion pipeline can run more than 30,000 compounds an hour. Members of the same team published subsequent work showing that further improvements of their deep fusion model could outperform either 3D-CNN or spatial graph CNN models alone, or biophysics-based techniques.[16]

Generative Models for De Novo Drug Design

The spillover of AI techniques into drug design and synthetic chemistry includes the use of generative models to suggest novel compounds for development. Cheminformatics tools for generative chemistry have existed for decades and serve as a baseline for judging performance of the new wave of AI-based generative models for exploring chemical space. These include MOSES,[17] GuacaMol,[18] and a new framework proposed by Zhang et al. in 2021.[19] The introduction of creative AI techniques around 2014 stemmed from the introduction of generative adversarial networks (GANs) by Ian Goodfellow, as well as variational autoencoders (VAEs) by Kigma and Welling, both of which have become popular architectures for generating highly realistic content, especially for images. The two main techniques work by developing a low-dimensional latent space of representations, whether it is images, video, speech, text, music, or molecular structures. These can then be sampled to build new creations.

GANs are built with two components, a generator and a discriminator, which compete against each other during model training. The generator builds a set of artificial data, and the discriminator attempts to distinguish it from real data. The generator module takes a latent point as input and then outputs an image (with pixel data). The model is trained until the discriminator is unable to distinguish the proposed "fake" data from the real data. For VAEs, the encoder performs the task of a generator. The encoded latent space with compressed representations can then be sampled with the decoder to map them back to molecular representations.

The first demonstrated use of GANs for molecule generation was in 2017, with objective-reinforced generative adversarial networks (ORGANs), followed later by an upgraded version called ORGANIC incorporating reinforcement learning. Generative models have been tested that combine GANs with autoencoders and GANs with RNNs.[20] Another proof point for the use of generative models emerged from a study that employed a deep LSTM network with transfer learning.[21] The experiment focused on discovery of agonists to modulate RXR and PPAR nuclear receptors, an important class of ligand-activated transcription factors. The study's investigators used a prior model that had been pre-trained with SMILES representations of 541,555 bioactive compounds selected from the ChEMBL20 database. A set of 25 fatty acid mimetics with known agonistic activity toward the two types of nuclear hormone receptors were then used for highly focused tuning of the model by transfer learning. The generated compounds were compu-tationally ranked by a QSAR model for PPAR and RXR activity. The five top-ranked compounds were synthesized and tested in bioassays. Two of the five were found to be selective for RXRs, two others were active against both receptor types, and one compound was reported inactive in test conditions. The study provided a clever early demonstration that generative models indeed could create potent agonists against these receptors. Along with the LSTM model, a string of positive results with other generative and deep learning approaches happened in parallel, with discoveries of inhibitors for the protein kinase JAK2 and optimized molecules for the D2 dopamine receptor.[22,23]

Some doubts and questions loom over the past reported attempts and future validation of generative models. Much of what these models create could be considered artifacts, as a good fraction of the molecules produced are not synthetically feasible, not drug-like, or with other features that suggest nothing was achieved other than a random generation of structures. Clarity around the composition of the training sets is also needed. For example, the 25 mimetics used for training in the final stage of the RXR/ PPAR receptor study were not disclosed. To pass scrutiny, training sets need to be published along with details of the underlying molecular generating protocol. The benchmarking protocols also suffer from lack of industry standards. A model might pass with flying colors under one benchmarking system but fail in another. A nice model for such a benchmarking framework was published by Bush and colleagues with the title "A Turing Test for Molecular Generators."[24]

A New Base of Innovation for the Pharmaceutical Industry

What we are trying to do with machine learning is not a task that people have ever been able to do very well and therefore we are not doing what I would call AI. We are applying learning techniques to large amounts of data to solve very hard problems.
— Daphne Koller, from *How machine learning is transforming drug discovery,*
January 2021

Biotechnology stands as the best model for how technology outside of small molecule chemistry R&D can expand the base of innovation in pharmaceutical research (Chapter 4). The biotechnology tools that spurred innovation were a natural fit, supplying techniques to investigate genes, study molecular mechanisms, and deliver biologic drugs. In contrast, AI and information technology seem worlds apart from the drug discovery apparatus. These closely related twins have matured over three successive generations with vastly improved techniques, computational power, and statistical methods. The pace of advancement in information technology is exponentially greater than in biology and medicine. Once deep learning's potential was proven by applications across many industries, venture capital poured in to chase the earliest trends for using AI in drug discovery. Fresh ideas and new methodologies from the mix of computing, data-driven biology and drug discovery seeded the first generation of startups scattered across the biotech hubs in the United States and Canada, and numerous AI-focused startups in China, with a few scattered throughout Europe. Table 7.1 presents a breakdown of some of the first startups and more recent noteworthy companies that have been using AI in drug discovery and development.

Table 7.1: Bringing AI Innovation to Drug Discovery

COMPANY	AI TECHNOLOGY APPLICATION	DRUG CANDIDATES OR BUSINESS DEALS	CAPITAL RAISED
Atomwise (2012)	Structure-based drug discovery	Multiple partnerships; collaboration with Lilly	$175M
Exscientia (2012)	Drug design	Multiple pharma partnership milestones; drug candidate for OCD	>$100M
BenevolentAI (2013)	Drug discovery	Partnership with AstraZeneca; phase 1 candidate for atopic dermatitis	$292M
Recursion (2013)	Phenotypic screening	Phase 2 trials planned for REC-4881, REC-3599, REC-2282, REC-994	$465M
Insilico Medicine (2014)	Drug discovery and design	Preclinical candidate for idiopathic pulmonary fibrosis; Pfizer collaboration	>$50M
XtalPi (2014)	Physics- and AI-based drug discovery	Drug discovery services	>$300M
Deep Genomics (2014)	Target and drug discovery	DG12P1 investigational new drug candidate	>$50M
Relay Therapeutics (2016)	Drug discovery and design	RLY-1971 partnered with Genentech, RLY-4008 and other early clinical candidates	Public (>$500M)
Cellarity (2017)	Target discovery and drug design	None disclosed	$173M

COMPANY	AI TECHNOLOGY APPLICATION	DRUG CANDIDATES OR BUSINESS DEALS	CAPITAL RAISED
Celsius Therapeutics (2018)	Target discovery	Collaboration agreement with Janssen; target discovery collaboration with Servier	$65M
Generate Biomedicines (2018)	Protein drug design	None disclosed	$50M
ConcertAI (2018)	Clinical trials and RWE	Drug development services	$150M
InSitro Therapeutics (2018)	Drug discovery	Discovery collaborations with BMS and Gilead	$743M
Valo Health (2019)	Drug discovery	Four targets disclosed	$400M
Genesis Therapeutics (2019)	Structure-based drug discovery	Collaboration with Genentech	$56M

Capital raised figures are current as of March 2021 and were discovered from publicly available (disclosed) information found in Pitchbook, Crunchbase, or company press releases.

Atomwise

Founded in 2012, Atomwise was a first mover in structure-based small molecule discovery with AI. The company developed its AtomNet technology using CNNs to predict the ability of small molecules to interact with protein targets. AtomNet models 3D protein structure together with compound structure, ultimately allowing the neural network to generalize to novel protein targets—those of high interest to pharma where there might not be any experimental biological activity data to rely upon for hit decision-making. Atomwise boasts the ability to screen billions of compounds, exploring chemical space beyond what is achievable in conventional pharma compound screens. The company has had considerable success in attracting partners in academic settings as well as in industry. Atomwise has announced major alliances with Merck (2015) and Lilly (2019) and is reported to have 285 partnerships in total. In 2020, Atomwise closed a $123 million dollar funding round and announced the formation of a joint venture with an Israeli biotech incubator, FutuRx. The new company, A2i Therapeutics, will use Atomwise's technology to screen 16 billion compounds for immune-oncology drug candidates.

Recursion Pharmaceuticals

Recursion has taken phenotypic screening and application of computer vision AI to unprecedented heights with its "Recursion Operating System" approach to biological target identification and drug hunting. According to its pre-IPO S-1 filing, the company has generated 8 petabytes of data from performing 66 million screening experiments in human cells. Those are astounding numbers for a young biotechnology startup. Like several of the first generation, AI-based drug discovery startups in Table 7.1, Recursion is tackling the discovery biology phase that is largely avoided by pharma. How have they fared? The sheer number of internal programs (37) and number of clinical stage candidates moving through Recursion's pipeline indicates that the company has built a robust discovery engine. This includes continued refinement of its AI, automation, and chemistry capabilities. For scaling its high throughput screening process, it is assembling compound libraries that *in toto* will soon approach the size of those in big pharma, containing more than one million compounds.

Recursion has built a stack of AI and cheminformatics tools to aid its blend of wet lab and *in silico* approaches. Critical components are those that enable deep filtering of candidates up front. Recursion's phenomics process utilizes tools to validate that the morphological phenotypes discovered by AI-based screening are indeed useful disease phenotypes for drug discovery purposes. Hits are then evaluated or predicted that can reverse the disease phenotype. A separate but related goal of the massive screening operation is the use of data to build a giga-scale catalog of inferred relationships. With each new experimental perturbation, such as a CRISPR knockout of a particular gene or the addition of a cellular growth factor, the company's platform can infer relationships between any two perturbations *in silico*. The continuous screening operation and inference engine has identified close to 100 billion relationships, a truly pioneering effort in the move toward *in silico* biology.

Deep Genomics

Founded by Brendan Frey from the University of Toronto in 2015, the startup is operating its AI Workbench to discover targets from analysis of genomics datasets. From published reports, the company's platform evaluates variants across the genomes or transcriptomes of normal and disease states and predicts those that might have therapeutic significance. In 2020, the company published results from a genomic analysis of Wilson disease, a rare metabolic disorder characterized by loss-of-function of a copper transporter leading to liver dysfunction and neurological disease.[25] The AI technology spotted an aberrant splicing mechanism arising as the result of a small variant in the sequence of the ATP7B gene. The discovery was uncovered by an AI model trained on splicing data, which together with other computational biology tools led to the *in silico* splicing prediction.

Relay Therapeutics

Relay has developed a molecular dynamics and protein modeling platform called Dynamo for small molecule drug discovery focused on precision oncology. The company has an agreement with D.E. Shaw Research to enable use of its Anton 2 supercomputer for long-timescale molecular dynamics simulations. The second-generation Anton 2 powers simulation of biomolecular systems with as many as 700,000 atoms. The essence of the platform is to model protein molecular motion, which is built to generate motion-based hypotheses for therapeutic design. The data from the simulations can predict novel binding sites and provides insights into strategies for modulating a protein's behavior.

A new breed of AI startups is bringing high-dimensional datasets from biology, mainly from genomics and imaging modalities, together with *active learning*, multitasking, or transfer learning approaches to marry computation and experimentation. Among the well-funded companies in this category are InSitro Therapeutics, Celsius, Cellarity, and Exscientia. Most, if not all, are employing human cell lines grown in a dish to screen drugs and model aspects of human disease. The use of active learning has become *de rigueur* as a means to overcome the limitations of small datasets with deep learning or the onerous labeling requirements for big datasets needed by supervised learning. More critically, active learning provides a feedback mechanism, evaluating data and then suggesting possible new experiments based on knowing what would improve the model; the next round then adds more data to the models. The knowledge comes from an active learning system that requests human annotation, or labeling, of instances with the most uncertain class assignment. The active learner can then produce a binary classifier based on smaller data while retaining good predictive performance. This iterative approach appears well-suited for discovery biology and early drug discovery work, where more informative experiments can be suggested as the machine learning model is being built.

Other notable AI-focused biotechnology startups outside of the group discussed are bringing new technologies to the fore of drug discovery.

Vijay Pande's laboratory at Stanford founded Genesis Therapeutics around protein structure prediction technology and has partnered early with Genentech. A few are using generative modeling to address some of the creative challenges in finding NCEs and druggable targets. These include Insilico Medicine, which reported the discovery of an inhibitor of DDR1, a kinase target for idiopathic pulmonary fibrosis in 2019.[26] Another company of note is Schrödinger, an established leader in physics-based methods but also combining AI for drug design. Companies building extensive biological knowledge graphs include Data4Cure (San Diego) and BenevolentAI (London and New York).

Partnerships across industries are playing a larger role for bringing in new computational ideas and AI software into the pharmaceutical industry. An MIT-led industry consortium—Machine Learning for Pharmaceutical Discovery and Synthesis—was formed in 2017 with a number of big pharma and biotech companies to develop software for drug discovery. Several tools were built or made available for synthesis planning (ASKCOS), molecular property prediction, and other AI applications for pharma. In 2019, Novartis struck a multiyear deal with Microsoft and is building an AI Innovation Lab with the tech giant. The German pharmaceutical company Boehringer Ingelheim announced a collaboration with Google in 2021 to explore quantum computing to accelerate drug discovery R&D. Bayer has a multiyear agreement with Schrödinger around drug design. NVIDIA and GSK announced an alliance around the AI chipmaker's expertise in GPU optimization and computational pipeline development, likely to bolster GSK's AI hub in London.

In many ways, these first-generation startups and hybrid partnerships are an important proving ground for AI-based drug discovery. There have yet to be clear indications that AI is moving the needle on biological target discovery or improving the efficiency of drug design in a way that biotechnology did with biologic drugs. Some industry observers and pharma executives remain skeptical at this juncture.[27] Judging from Pharma's low level of acquisition activity around AI, the applications or technology platforms have not provided compelling proof points of value. In drug development, the value inflection points are demonstrated only after a drug candidate clears the hurdles faced in human clinical trials. No "AI drugs" have made it through Phase 2 or Phase 3 trials since the appearance of the first class of AI startups, which were founded in 2012–2014. Thus, there is no poster child for a miracle AI drug as was the case for immunotherapies such as Yervoy, Keytruda, and Opdivo, or other innovative barrier-breakers.

Summary

Drug discovery embraces a huge swath of scientific methods sampled from chemistry, biophysics, quantum mechanics, physiology, genomics, molecular biology, proteomics, pharmacology, AI and machine learning, and information theory. Pharmaceutical industry teams have in some instances been at the forefront of exploring modern AI methods to augment or replace cheminformatics techniques that have driven novel compound exploration. New biotechnology startups are pushing the boundaries of integrating AI with target discovery, virtual screening, and introducing generative drug design methods to complement big pharma's efforts. The feature here highlights the expanded utilization of AI across industry in the early phases of drug discovery within the past decade.

UTILIZATION OF AI TECHNOLOGY ACROSS DRUG DISCOVERY AND DESIGN

- Predicting chemical properties (ML, DL)
- Learning molecular representations (ML, DL)
- Optimizing chemical reactions (RL)
- Predicting organic reaction outcomes (NLP, Symbolic AI + NNs)
- *De novo* drug design (RNNs + RL, GANs, VAEs)
- Generating novel targeted chemical libraries (RL, GANs)
- Predicting compound toxicity (ensemble ML)
- Modeling ADMET endpoints (multi-task graph CNNs, one-shot learning)
- Solving 3D protein structures (DL)
- Models for ligand-based virtual screening (multi-task DNNs)
- DMTA cycle optimization (active learning)
- Guiding multi-objective optimization (ML/DL)

Area of application is followed in parentheses by AI method(s) being investigated across the pharmaceutical industry and in AI-based drug discovery startups.

A final thought relates to AI and automation. Across biological and chemical laboratories in industry, AI and machine learning are expected to play a key role in automation performance, resulting in faster adoption of new synthesis methods and facilitating the synthesis of a more extensive range of compounds from chemical space. The examples from this chapter highlight how synthetic chemistry can be taught to machines using deep learning, using new AI architectures and high-quality datasets. Integrating these tools into production will require patience and work to overcome cultural and practical hurdles that hinder fully automated processes. In addition, complete automation of the DMTA cycle has not yet been achieved but will benefit from efforts within industry and across small biotechnology startups to industrialize active learning methodologies.

Success with the iterative process linking computational predictions with lab experimentation would alter thinking around the discovery path in pharma. Foremost, it might force a closer look at bringing more resources to tackle biological problems that could lead to new ways to intervene in diseases, find druggable targets, and increase efficiency.

Pharma benefits from the amazing proliferation and maturation of AI approaches developed in the technology sphere, ensuring that an arsenal of tools will become available for scientific discovery. These information-based technologies are poised to transform medicinal chemistry and one day may become responsible for autonomously directing the DMTA cycle as a smart factory, after tight integration with robotics and automation. The laboratory of the future is moving into the age of AI.

Notes

1. Corey EJ, Wipke WT. Computer-Assisted Design of Complex Organic Syntheses. Science. 1969 Oct 10;166(3902):178–92.

2. Lee I-D, Palsson BO. A Macintosh software package for simulation of human red blood cell metabolism. Comput Methods Programs Biomed. 1992 Aug 1;38(4):195–226.

3. Mullard A. Supersized virtual screening offers potent leads. Nat Rev Drug Discov. 2019 Mar 5;18(243).

4. Lipinski CA, Lombardo F, Dominy BW, Feeney PJ. Experimental and computational approaches to estimate solubility and permeability in drug discovery and development settings. Adv Drug Deliv Rev. 1997 Jan 15;23(1):3–25.

5. Lloyd DG, Golfis G, Knox AJS, Fayne D, Meegan MJ, Oprea TI. Oncology exploration: charting cancer medicinal chemistry space. Drug Discov Today. 2006 Feb;11(3–4):149–59.

6. Jorgensen WL. The Many Roles of Computation in Drug Discovery. Science. 2004 Mar 19;303(5665):1813–8.

7. Lyu J, Wang S, Balius TE, Singh I, Levit A, Moroz YS, et al. Ultra-large library docking for discovering new chemotypes. Nature. 2019 Feb;566(7743):224–9.

8. Brown DG, Boström J. Analysis of Past and Present Synthetic Methodologies on Medicinal Chemistry: Where Have All the New Reactions Gone? J Med Chem. 2016 May 26;59(10):4443–58.

9. Segler MHS, Preuss M, Waller MP. Planning chemical syntheses with deep neural networks and symbolic AI. Nature. 2018 Mar;555(7698):604–10.

10. Segler MHS, Waller MP. Modelling Chemical Reasoning to Predict and Invent Reactions. Chem – Eur J. 2017;23(25):6118–28.

11. Coley CW, Barzilay R, Jaakkola TS, Green WH, Jensen KF. Prediction of Organic Reaction Outcomes Using Machine Learning. ACS Cent Sci. 2017 May 24;3(5):434–43.

12. Schwaller P, Gaudin T, Lányi D, Bekas C, Laino T. "Found in Translation": predicting outcomes of complex organic chemistry reactions using neural sequence-to-sequence models. Chem Sci. 2018 Jul 18;9(28):6091–8.

13. Tetko IV, Karpov P, Van Deursen R, Godin G. State-of-the-art augmented NLP transformer models for direct and single-step retrosynthesis. Nat Commun. 2020 Nov 4;11(1):5575.

14. Schwaller P, Hoover B, Reymond J-L, Strobelt H, Laino T. Extraction of organic chemistry grammar from unsupervised learning of chemical reactions. Sci Adv. 2021 Apr;7(15):eabe4166.

15. Stevenson GA, Jones D, Kim H, Bennett WFD, Bennion BJ, Borucki M, et al. High-Throughput Virtual Screening of Small Molecule Inhibitors for SARS-CoV-2 Protein Targets with Deep Fusion Models. ArXiv210404547 Cs Q-Bio. 2021 Apr 9.

16. Jones D, Kim H, Zhang X, Zemla A, Stevenson G, Bennett WFD, et al. Improved Protein–Ligand Binding Affinity Prediction with Structure-Based Deep Fusion Inference. J Chem Inf Model. 2021 Apr 26;61(4):1583–92.

17. Polykovskiy D, Zhebrak A, Sanchez-Lengeling B, Golovanov S, Tatanov O, Belyaev S, et al. Molecular Sets (MOSES): A Benchmarking Platform for Molecular Generation Models. Front Pharmacol. 2020 Dec 18.

18. Brown N, Fiscato M, Segler MHS, Vaucher AC. GuacaMol: Benchmarking Models for de Novo Molecular Design. J Chem Inf Model. 2019 Mar 25;59(3):1096–108.

19. Zhang J, Mercado R, Engkvist O, Chen H. Comparative Study of Deep Generative Models on Chemical Space Coverage (v18). 2021 Feb. Available from: `https://chemrxiv.org/articles/preprint/ Comparative_Study_of_Deep_Generative_Models_on_Chemical_ Space_Coverage/13234289/3`.

20. Putin E, Asadulaev A, Ivanenkov Y, Aladinskiy V, Sanchez-Lengeling B, Aspuru-Guzik A, et al. Reinforced Adversarial Neural Computer for de Novo Molecular Design. J Chem Inf Model. 2018 Jun 25;58(6):1194–204.

21. Merk D, Friedrich L, Grisoni F, Schneider G. De Novo Design of Bioactive Small Molecules by Artificial Intelligence. Mol Inform. 2018;37(1–2):1700153.

22. Popova M, Isayev O, Tropsha A. Deep reinforcement learning for de novo drug design. Sci Adv. 2018 Jul 25;4(7).

23. Olivecrona M, Blaschke T, Engkvist O, Chen H. Molecular de-novo design through deep reinforcement learning. J Cheminformatics. 2017 Sep 4.

24. Bush JT, Pogany P, Pickett SD, Barker M, Baxter A, Campos S, et al. A Turing Test for Molecular Generators. J Med Chem. 2020;8.

25. Merico D, Spickett C, O'Hara M, Kakaradov B, Deshwar AG, Fradkin P, et al. ATP7B variant c.1934T > G p.Met645Arg causes Wilson disease by promoting exon 6 skipping. Npj Genomic Med. 2020 Apr 8;5(1):1–7.

26. Zhavoronkov A, Ivanenkov YA, Aliper A, Veselov MS, Aladinskiy VA, Aladinskaya AV, et al. Deep learning enables rapid identification of potent DDR1 kinase inhibitors. Nat Biotechnol. 2019;37(9):1038–40.

27. Lowe D. AI and Drug Discovery: Attacking the Right Problems. In the Pipeline. 2021. Available from: `https://blogs.sciencemag. org/pipeline/archives/2021/03/19/ai-and-drug-discovery- attacking-the-right-problems`.

CHAPTER 8

Biotechnology, AI, and Medicine's Future

The task is not so much to see what no one has yet seen, but to think what nobody has yet thought about that which everybody sees.

—Erwin Schrödinger

We walk backward into the future, with our minds anchored by the imprinted lessons of the past.

—Wayne Wagner

Forward thinking in the midst of rapid technological change is difficult; for some, it is obvious where the world is headed, while to others, the path is blanketed by a mysterious fog. Given AI and machine learning's penetration into science and society, and the misunderstanding that often surrounds its myriad applications, fog is an apt metaphor for AI. By nearly all accounts, AI is at the forefront as a driver of the expanding and deepening digital revolution. In his book *The Fourth Industrial Revolution* (Crown Business, 2016), Klaus Schwab casts the current industrial revolution as a function of the synergies found in connecting a more ubiquitous and mobile Internet together with sensors and AI/ML technologies.[1] He cites megatrends in the physical, digital, and biological realms as the key drivers from an economic standpoint. The world is undergoing physical change with the appearance of autonomous vehicles, robotics, and 3D printing technology, all made possible to a large extent by AI and advances in engineering. Sensors and digital technologies are connecting the globe as never before. In biology, the trio of genetic engineering, genomics, and neurotechnology are pushing the frontiers of bioengineering and medicine. From the viewpoint of medicine, innovations in biotechnology and computer science (hardware and software, now with AI) are top trends.

Since the start of modern science and technology post–World War II, computer science and biological sciences have been converging to form new disciplines such as computational biology. Underpinning modern trends and industrial revolutions are the groundswells of new thought and technology breakthroughs that fuel paradigm shifts within disciplines. The convergence of disciplines can lead to paradigm shifts or unifying theories that occur over different timeframes, such as with the sudden emergence of evolutionary theory out of the convergence of biology, geology, and paleontology; the rapid development of the quantum theory of the atom and the nature of chemical bonds (physics and chemistry); or the formulation and widespread acceptance of the "central dogma" and information flow in biology (molecular biology and chemistry).

The observation that important fields of science and engineering are converging to impact healthcare and the life sciences is not new; it was described by an MIT report in 2011. The convergence of life sciences, physical sciences, and engineering was anticipated to bring about rapid innovation and advances in medicine as well as many fields outside of medicine. This coalescence is not simply a matter of interdisciplinary work or special projects at a defined intersection. In several domains of biology, it would be impossible to remove the data science and algorithmic approaches necessary for the science to be done, much of which has been explored in this book.

A generation of new technologies, algorithms, and architectures are taking hold and transforming the innovation landscape. Technologists and venture capitalists active within healthcare have already spied many opportunities to develop medicines driven by the twin engines of AI and biotechnology. They have witnessed inflection points in value obtained through successes in modeling and simulation, the wide availability of pharmaceutical data, clinical trial results, use of population-scale genomic information, and precision molecular engineering.

One of the biggest impacts of the tech revolution on the life sciences is the tools that it delivers for accelerating cycles of discovery, as pictured in Figure 8.1. Hypothesis generation is essential across biology and drug discovery, as it is for most of exploratory science. Testable hypotheses are evaluated with carefully designed laboratory experiments, or with computational methods, using platforms that generate data. In medicine, analysis of disease conditions and definitive diagnoses are reached by logical deduction, with inclusion or dismissal of disease hypotheses during the differential diagnosis process. The diagnostic procedure in medicine is augmented with testing and data generation. In biology, pharma, and medicine, the initial data processing pipelines come increasingly from the AI/ML data science toolkit. Insights are derived from secondary analysis tools built for computational biology or chemistry or medical diagnostics. The application of AI methods is often the driver for hypothesis-free investigations that search for patterns or attempt to find associations in datasets. Generating and testing hypotheses promises to be a main benefit of AI. Since testing and running technology platforms is often costly, computational methods to generate and evaluate hypotheses reliably becomes critical. The innovative tools of biotechnology are crafted from insights taken from each of these spheres. The industry leans heavily on the ability of AI/ML methods to optimize engineering platforms and aid in the discovery and prioritization of new therapeutics.

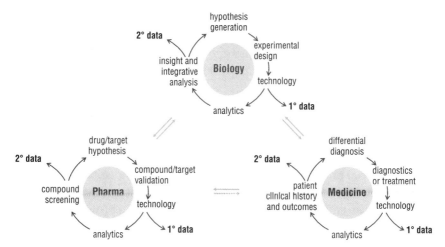

Figure 8.1: The discovery cycles in biology, pharma, and medicine

The process of generating biological insights, discovering new pharmaceutical compounds, and practicing medicine is depicted in three separate but related cycles. Insights from biology, such as disease mechanisms, translate into new drug hypotheses. Testing of novel drugs and clinical trials informs medical practice. Patient outcomes and clinical histories aid in refining new hypotheses about disease treatments and human biology. Data science is critical for the analysis of primary and secondary data from these cycles.

Building Tools to Decipher Molecular Structures and Biological Systems

During the past fifteen years we have been attacking the problem of the structure of proteins in several ways. One of these ways is the complete and accurate determination of the crystal structure of amino acids, peptides, and other simple substances related to proteins, in order that information about interatomic distances, bond angles, and other configurational parameters might be obtained that would permit the reliable prediction of reasonable configurations for the polypeptide chain.[2]

—Linus Pauling, Robert B. Corey, and H.R. Branson, 1951

At the heart of biotechnology and drug discovery programs are the molecular structures of life. Unraveling the mysteries of the cell's information flow started with solving the structure of DNA, determining the genetic code, and demonstrating its universality. These steps (reviewed in Chapter 4) were only the start of a process that continues today to describe the organization, structure, and function of both the genome and the proteome. It is now widely appreciated that the higher-order structure of chromosomes and epigenetic modifications encode additional information to regulate gene expression and control DNA replication and cell division. Several technologies have been developed over the past decade to detect and quantify 3D conformations, megabase-scale domains, and regions of chromatin accessibility. More recently, machine learning approaches have been applied to predict these structures and the DNA sequence locations of chromosomal modifications.

Two of the most important computational challenges in biology now appear to be within reach with AI and machine learning. The first is the long-sought goal of enabling prediction of the atomic-level 3D structure of any protein from its primary (1D) sequence of amino acids. The accurate computational prediction based on sequence strings alone, without needing protein crystallography, would greatly impact biological research. Evolution has created hundreds of millions of unique proteins found in tens of millions of organisms. Structural details would immediately suggest or support functional predictions from a gene or protein sequence, a second grand challenge in biology. For protein function elucidation, primary sequence may not be sufficient, and predictive methods may require some knowledge of the protein's ultimate microenvironment. One case would be where a discovered protein belongs to a large multiprotein complex. An example is mitochondrial complex I, a molecular machine containing 45 subunits that carries out electron transport and generates a proton gradient across the inner mitochondrial membrane. A subunit in this complex may not have an obvious functional domain or a catalytic site and may serve an accessory structural role in the complex.

Formulating accurate 3D structural details and cataloging the functions of nearly all proteins will be a fundamental advance for biology. For applications outside of medicine, engineering of new enzymes and biocatalytic processes should scale rapidly. However, in mammalian systems, additional information will be needed to understand the rules of transcriptional regulation and cellular functions that are dictated to a large degree by DNA-protein and protein-protein interaction cascades, respectively. This layer of molecular organization is important for determining the mechanism of action of drugs and the pathophysiology of disease. Analysis of such interactions has been facilitated by deep learning, whether for predicting transcription factor binding, DNA methylation, translation initiation sites, or for inferring locations of enhancers and insulators along the genome. Similarly, AI has been used to discover patterns of regulatory RNA binding sites along the genome and to predict RNA expression patterns in tissues.

At the next level of organization, gene or protein interaction networks dictate or strongly modulate cellular responses and cell states under normal or disease conditions. Network analysis is therefore another research area where the use of pattern recognition algorithms from AI will yield important insights into cellular behavior. Beyond intracellular networks, tracking cell-cell communication, recognizing protein interaction networks, and finding neuronal connectivity patterns are all targetable by AI methodologies.

For unlocking insights across the biological spectrum, "omics" technologies used for molecular profiling are currently the best tools for generating the mountains of data needed to fuel machine learning algorithms at scale. These include genomics sequencing-based applications that reveal chromatin accessibility (ATACseq), single cell resolution of transcription (single cell RNA sequencing, or scRNAseq), spatial transcriptomics for localizing hundreds to thousands of RNAs, and mass spectrometry and antibody or oligonucleotide array-based methods to define the proteome. Nearly all of these methods require computational biology methods in tandem with the technology platforms. Figure 8.2 shows a breakdown of the current technology stack for biology.

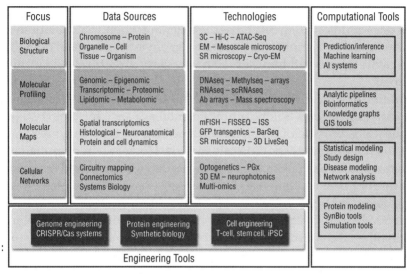

Figure 8.2: The tech stack for biology

Modern approaches to interrogate biology combine an array of technologies, experimental and engineering frameworks, and computational methods. Biological structure analysis is performed at the level of individual molecules to entire organisms with different types of microscopy and image analytics. Molecular profiling is now attainable for all of the "omics" using arrays and high throughput instrumentation. Spatial profiling and connectivity mapping are now enabled with increasing resolution by instrumentation and AI algorithms. Engineering tools enabling gene and cell manipulation, or perturbation experiments, are now routinely incorporating statistical methods and algorithms for optimization.

GIS: geospatial information science

SR: super-resolution

AlphaFold: Going Deep in Protein Structure Prediction

A cornerstone achievement of AI for biology was announced by DeepMind in November 2020.[3] The company's AlphaFold program vaulted into new territory by solving protein structures using its updated AI system that had never been achieved previously without precise crystallographic or cryo-EM data. Upon arrival of the news, the AI and scientific research communities branded the new AlphaFold as one of the top two breakthroughs in AI for 2020. The earlier version of the program had predicted several 3D structures from the sequence of the Sars-CoV-2 genome (see Chapter 1). The team at DeepMind reengineered the AI system, and its performance characteristics were made public after release of the results from the 14th CASP competition. The DeepMind approach that led to the startling performance was designed with a modular neural network architecture that allows the system to derive structural information from evolutionarily related protein sequences and amino acid residue pairs in an iterative learning process. The AlphaFold system obtained a median score of 92.4 on the Global Distance Test, a measure of the percentage of amino acids found within atomic range (1.6 Ångstroms) of precise coordinates for the correct structure. The striking performance gains by the DeepMind team over the years in the free modeling (without use of structural templates) competition underscores the pace of AI technology development at the company. The best performing teams from 2006–2016 reached median scores of only 40; AlphaFold in 2018 scored 58 and AlphaFold 2 in 2020 attained 87, far outpacing the competition.

With these results, the scientific world awaits publication of the details of how AlphaFold has solved one of the most important and difficult computational challenges in biology. It may well be that this is one of those problems that is uniquely well suited to being solved by neural network architectures, given the immense amount of sequence data available to the algorithm and the model's capacity. Real-world applications in industry would be enabled by this technology, as the accuracy is sufficient to determine catalytic mechanisms of enzymes and drug-target interaction sites. Scientists in academia will also want to know whether this resource will be made available for research, and at what cost, since the developers are part of an enterprise controlled by Google's parent company, Alphabet.

The dramatic improvements in structure solving did not come out of the blue. The two main components of AlphaFold's approach were introduced to the field decades ago. The first is the use of co-evolutionary relationships of amino acid pairs that are in close physical proximity, which can be converted to a binary contact map to encode the spatial information necessary for folding. The second component evaluates multiple sequence alignments of related sequences to extract rules and learn evolutionary constraints. The AlphaFold system was trained with both 3D protein structures out of PDB and enormous sequence databases containing hundreds of millions of sequences.

Deep learning models have now surpassed all other methods for performing a central task for structure analysis—protein contact prediction. This has been achieved with supervised contact prediction with deep residual networks trained on many protein structures,[4] as well as with unsupervised methods such as GREMLIN trained on sequence data without structural information.[5] Tremendous progress has also been made with the use of protein language modeling using transformers and other approaches for contact prediction. The aim of these efforts has been to determine structural information to inform protein design, rather than build accurate structural hypotheses.[6-8]

Predicting Genome 3D Organization and Regulatory Elements

The unveiling of the hierarchical organization of the genome in 3D space over the past decade is one of the most fascinating and fast-moving areas in the field of genomics. Its importance in biology is tied primarily to genome integrity, gene regulation, and DNA replication. Chromatin structure is studied experimentally with imaging techniques based on fluorescence in situ hybridization (FISH) or super resolution microscopy, and with molecular methods that are either proximity ligation-based (Hi-C, scHi-C, DNA-FISH) or ligation-free (ChIA-Drop, GAM, SPRITE). The latter are categorized as chromosome conformation capture (3C) techniques and have generated insights over topologies and domains spanning from tens of nucleotides to megabases in scale. Important regulatory features are promoter-enhancer pairs, chromatin loops, topologically associated domains (TADs) of tens to hundreds of kilobases, and long-range TAD-TAD contacts spanning megabases of chromosomal DNA.

As is the case with protein folding and 3D structure problems, the prediction of contact points is critical to defining the sequence determinants that drive chromosome conformation and overall nuclear organization. Deep learning again has proved pivotal in providing systems that can predict contact points using DNA sequence alone, or with epigenomic or derived features from genome folding as inputs. Tools recently developed and in use for predicting contacts are CNN-based Akita[9] for many types of contact predictions, DeepTACT[10] for inferring promoter-enhancer contacts, DeepMILO[11] for noncoding variant impact on folding, and DeepC[12] for megascale boundary prediction.

AI algorithms are now largely eclipsing techniques for detection or prediction of regulatory elements, DNA-protein binding sites, and epigenetically modified regions across the genome. Some of these deep learning applications can detect more than one class, such as DanQ, which reads DNA sequence and predicts transcription factor binding, histone modifications, and DNase hypersensitivity sites.[13] Another is DeepBind, a software application built to harbor several models trained to predict deleterious sequence variants, transcription factor motifs, and RNA-binding protein binding sites—important in alternative splice prediction—using data from a variety of high-throughput sequencing assays.[14] Deep learning systems are bringing *in silico* predictions further into molecular biology research than ever before.

AI Approaches to Link Genetics-Based Targets to Disease

The search for genes linked to disease and modifier genes that could be used as guides to alter disease progression or severity has been guided over the past century by genetic studies in model organisms, such as fruit flies and mice, and in humans. The use of GWAS and human cohorts as a basis for disease gene discovery has benefited in the past two decades from the ability to scale studies from hundreds of genomes to large-scale efforts surveying SNPs from more than 100,000 individuals. While several of these landmark research investigations are a celebration of scientific collaboration, they have yielded fewer than expected leads for therapeutic development. To address this, computational biologists have again turned to deep learning approaches for identifying patterns and signals that would lead drug developers to disease-modifying pathways or molecular targets.

The power of genetics comes from nature's experiments across individuals in populations. Family-based studies are needed in human genetics to lower the genetic background differences found across populations and also to pinpoint inherited mutations that are causative of disease. Olga Troyanskaya and Richard Darnell led a genetics and AI-based study on autism spectrum disorder (ASD) in 2019 that illustrates the value of deep learning tied to massive genomic sequence data.[15] The research strategy employed deep learning on whole-genome sequencing data from 7,097 genomes in 1,790 families, each with one affected member, to find genetic regions linked to the disorder. The appearance of the disorder in these families is due to *de novo* mutations passed on from the unaffected parents to offspring. Newborns typically inherit fewer than 100 *de novo* mutations and are difficult to spot with sequencing technologies, as their frequency ($1x10^{-8}$) is much lower than sequencing error rates. However, with family information, predictions of *de novo* variants are more reliable as the parental alleles are present for the variant calling algorithm to utilize.

A deep learning framework was developed to predict the effects and functional consequences of deleterious genetic variants occurring in nonprotein coding regions, as the vast majority of *de novo* variants occur in this vast "dark matter" of the genome. The deep learning CNN-based model deployed is an improved version of DeepSEA[16] and was trained on multiple datasets for DNA and RNA binding proteins and their sequence targets—utilizing more than 2,000 features. For each genome in the study, they predicted potential functional impact of *de novo* variants and then identified mutations that disrupted either transcriptional or post-transcriptional regulation. The results were very impressive on several fronts. Analysis of the noncoding *de novo* variants in affected individuals suggested that both transcriptional and post-transcriptional dysregulation may play an important role in manifestation of known ASD disease phenotypes. The most remarkable aspect of the data was the demonstration that these noncoding mutations converge on genes with roles in brain-specific functions (synaptic components) and neurodevelopmental processes tied to ASD and other neurodevelopmental disorders.

Using human genomes or artificially derived genomes as a starting point, researchers have designed AI systems to perform *in silico* mutagenesis to predict the location and identities of sequence variants that might drive disease processes. For example, to predict the effect of genetic variants, AI models can be trained against the reference genome datasets and *in silico* mutagenized sequence. GANs or other models can then discriminate differences of relevance to the disease phenotype. Alternatively, predictions on functional impact can be predicted separately from normal versus disease, and differences compared and assessed as candidate disease-related variations.

Quantum Computing for In Silico Chemistry and Biology

The growing body of knowledge around protein 3D structure with new computational tools and the relationship between structure, function, and sequence should produce a leap forward in determining novel protein function and assessments of drug targetability. A main task in the field of bioinformatics had historically been about building models or using sequence similarity search tools for predicting an uncharacterized gene's encoded protein function. Less emphasis has been placed on modeling molecular processes or cellular phenomena, since the modeling effort requires large computing resources available historically to very few laboratories. Developing realistic biological system models and chemical quantum mechanical simulations has been slowed by both lack of knowledge on the biology side and compute power for chemistry. To tackle these challenges, quantum computing looks promising, if not necessary, to perform *in silico* work and model these complex, probabilistic systems with quantum logic and the requisite vast number of calculations.

Today's quantum computers are still in their infancy with limited practical applicability to real-world problems. Quantum computing is a new paradigm, replacing standard binary (0 and 1) "bit" computing with qubits, which allow for state superposition (both 0 and 1). The first systems built arrived with under 100 physical qubits. Google's Bristlecone is 72 qubits, Righetti's Aspen-9 has 31, and IBM and Honeywell have a series of machines with similar processor power. IBM's roadmap includes the 1,121-qubit Condor scheduled for release in 2023. These machines use superconducting quantum processors and special quantum computing algorithms. Tools for quantum computing are available from Zapata Computing, IBM, Cambridge Quantum Computing, Microsoft, Google, and others.

Estimates vary as to the quantum computing threshold needed for molecular modeling and biological dynamic systems. It is likely that quantum computers will need 1,000 qubits before they can usefully model chemical systems. Most of the research work done now is exploratory, with hopes to plan for proof-of-concept testing on small applications. First, fundamental engineering challenges need to be solved. Getting to a quantum transistor with logic qubits performing quantum operations together will require development and testing to demonstrate the encoding of one logical qubit from 1,000 physical qubits. At that point, when qubits are interacting with each other in performing a computation, the power of the quantum computer will become evident. With superposition error-correcting qubits, quantum algorithm experts will have available an exponential expansion in the number of values that can all be considered at the same time for number crunching on the most intensive problems.

Neuroscience and AI: Modeling Brain and Behavior

How is vision, which we do effortlessly, transformed into action, which requires volition, which requires attention and engagement? How this transformation takes place is a fundamental question that is at the heart of brain function and behavior.

—Mriganka Sur, Newton professor of neuroscience at MIT, in an interview with STAT, July 2018

The human brain is often thought of as a decision-making machine; it is also a multimodal perception engine with enormous plasticity geared toward learning and memory—all with immense, prewired (innate) cognitive capabilities. Vast improvements have been made over the past 50 years for describing the details of an individual neuron's activity, for measuring responses in small neuronal circuits, and more recently for monitoring large networks. Remarkable technology is now available to control circuits experimentally and to visualize networks with high precision. Despite these technical advances and accumulated data, modern neuroscience still does not have a theory for how the brain works, nor a plausible route to get there. It is unlikely to do so in the absence of a new computational framework.

Why are frameworks and models important in understanding the workings of the brain? The answer comes from the beginnings of modern neuroscience and the unifying power of joining experimental observations with theory. The Hodgkin-Huxley model of neuronal excitability provided exceptional explanatory power for understanding single neurons and a basis for constructing theories of circuits. The duo published a series of five papers in the *Journal of Physiology* in 1952 that brought together experimental findings with a quantitative model found in their famous equations. The macroscopic understanding of electrical currents and the action potential led to a search for the underlying molecular components—ionic channels and their regulators. Combined with Hebb's rule of learning, known informally by the phrase "neurons that wire together, fire together," a small set of principles dominated the construction of models for how neurons compute and a logic to their wiring diagrams and circuit operations.

Systems neurobiologists look beyond individual neuron response properties to include analysis of groups of neurons in small circuits, such as in reflex arcs, where computations can be modeled together to predict circuit function and animal behavior. There are now well-described circuits for pain processing, memory consolidation, hypothalamic regulation of energy and glucose homeostasis, and complex pathways intersecting at the basal ganglia for sensorimotor control and decision-making. Quantitative modeling of circuitry has encountered difficulty with scaling, and at present there are no satisfying models of emergent computation in the brain.

Computational neuroscientists and AI researchers have long flirted with artificial neural networks in attempts to build models of the brain. Progress was slow, and enthusiasm stalled up until artificial neural networks (ANNs) took off in the 2000s with far more computational muscle available from GPUs and the arrival of big data. Proponents of reintroducing the use of ANNs for modeling the brain pointed to success in computer vision and the observation that deep layers in hierarchical architectures derived representation spaces that strongly resembled those from abstractions present in higher visual areas of the primate brain. In 2015, Nikolaus Kriegeskorte outlined an idea of a technical framework from which to begin the work.[17] Kriegeskorte argued that systems neuroscientists needed to move away from classical computational models, where use of shallow architectures and simple computations in modeling information processing in the brain were limited in power. As he phrased it: "We are now in a position to integrate neural network theory with empirical systems neuroscience and to build models that engage the complexities of real-world tasks, use biologically plausible computational mechanisms, and predict neurophysiological and behavioral data." Although Kriegeskorte focused his framework on biological vision and information processing, the thrust of his idea—that computational models using feedforward and recurrent neural networks could model the brain —would be embraced and refined in a second iteration.

AI has made a huge resurgence in neuroscience research, with studies showing that deep learning in particular can serve as a basis for modeling a variety of brain functions or systems. These include motor control, reward prediction, audition, vision, navigation, and cognitive control. The continued string of successes and AI research into deep learning methods has led to the uncovering of a set of principles that are key to its stellar performance on a set of narrow tasks in the "AI set" versus those found in the brain set.[18] Success of ANNs has been predicated on three components: human design and selection of network architectures, learning rules, and objective functions. A new proposal outlined by a large contingent of neuroscientists and AI researchers argues for an optimization-based framework leveraging these core features of ANNs.[19]

One of the most compelling aspects of this framework is the tantalizing prospect that AI models might be used as a guide to understanding emergent computation in the brain. Deep neural networks already display this property. No current theories of the brain adequately account for this important phenomenon. A test of the framework would be a demonstration that neural responses are an "emergent consequence of the interplay between objective functions, architecture and learning rules."[19] Placing an emphasis on objective functions raises the question as to their observability in the brain. The authors argue that even though some objective functions may not be directly observable, they can be defined mathematically without necessarily referring to a specific task. More straightforward are the learning rules, which are the recipes for updating synaptic weights in a network. The architecture in the brain is the collection of pathways and hierarchical connections for information flow. Thus, one key to developing models will be well-defined circuitry maps of the systems (for example, feeding behavior or fine motor control) under question. A big question is whether enough is known about neuroanatomical circuits and how they behave. The most important readouts of this framework will be the accurate prediction of behavioral measurements, such as stimulus-response times, motor task activities, visual pattern detection, and the shape or trajectory of data from continuous tasks.

Modeling the brain based on artificial learning systems has long drawn criticism, primarily because of the observation that an ANN has little resemblance to brain networks and the stark difference in how neurons in an artificial network compute versus those in a nervous system. It is worth noting that neuroscience has a long history of philosophers, psychologists, and neuroscientists dismissing new ideas on nervous system function that later became foundational elements for the discipline. In his book *The Idea of the Brain: The Past and Future of Neuroscience* (Basic Books, 2020), Matthew Cobb paints a picture of some of this history, starting with the widely held pre-Renaissance view that the seat of thought and consciousness was to be found in the heart, not the head.[20] Not until the end of the 18th century did scientific knowledge begin to have an impact on the strong philosophical and religious objections against materialist explanations of the mind. Prior to the 20th century, the notion that the nervous system worked by electrical signaling was dismissed or considered controversial, as was the concept that there were separate cells that communicated with each other in brain tissue. Chemical neurotransmission was not believed to have a role in brain activity until well into the 20th century. By the mid-20th century, two amino acids were discovered to be active in the brain—GABA and glutamate—but were nevertheless dismissed as being important chemical synaptic mediators. These were only accepted as conforming to the classical criteria of neurotransmitters in the late 1960s and 1970s. The list goes on; genes and transcriptional regulation were thought to have little or no role in synaptic function as late as 1990. The most recent fad is to push back against "biologically implausible" algorithms that could not possibly be implemented in the brain, such as backpropagation or XOR logic in single neurons. The ability of AI to inform neuroscience and decode neural network operations is on track to be another fascinating and controversial chapter.

Brain Information Processing and Modularity: Climbing a Granite Wall

Demis Hassabis, the CEO and co-founder of DeepMind, is also an avant-garde AI researcher pursuing big ideas in neuroscience. He and his academic colleagues in the United Kingdom envision a new era of collaboration between AI researchers and neuroscientists, a "virtuous circle" whereby neuroscience gains by learning from AI, and vice versa.[21] In the past, AI research has borrowed considerably from the neurosciences, producing the deep neural networks inspired by the brain's hierarchical architecture for visual information processing and powerful algorithms based on reinforcement learning, temporal difference learning, and others. How will neuroscience now benefit from AI? Insights from new learning algorithms in AI may spark a search for similar functionality in the brain. The best example of this is the biological correlate of the backpropagation algorithm. Independent studies have now supported the idea that a type of backpropagation method can indeed operate in the brain. Hassabis and colleagues have steered clear of offering up a bold modeling framework around the principles of learning systems derived from ANNs and instead reiterate their faith in the modular construction and operation of the brain. They believe that imposing an AI-oriented framework focused on a global optimization principle whereby the brain learns with a single, uniform network misses the point of its modular organization and interactive nature.

A fun example of the modular operation of the brain and how it might be modeled comes from extreme sports: Alex Honnold's free soloing of the massive granite wall of El Capitan in Yosemite National Park on June 3, 2017. If one stops to think about all the cognitive and sensorimotor capabilities that are needed to complete such an endeavor without ropes—beyond the obvious nerves of steel and sheer athleticism required—a clear picture of what types of brain modules are likely to be engaged emerges. First, the task being modeled is route completion, much like driving an autonomous vehicle in a race around the city or across the desert to reach a finish line. In Alex's case, the mapped route is called Freerider, which zigzags and traverses 3,000 feet up the vertical rock face with hundreds, if not thousands, of positional markers, toe holds, and finger ledges.

Getting to the top alive for an elite rock climber is not just a matter of careful action planning, motor control, sensory perception, and focus to complete a one-off attempt. For Alex's feat, successful real-time execution entailed years of memorization of the route's intricate details, learning maneuvers with sensorimotor feedback, building physical adaptations for motion control, and increasing psychological preparedness. Honnold's experience in free soloing undoubtedly tuned his brain's systems to accomplish such feats. He had previously soloed some of the most difficult technical routes on the planet, including the Rainbow Wall in Red Rocks, Nevada (graded V 5.12a and covers 1,150 feet), the Crucifix on Higher Cathedral Rock in Yosemite (IV 5.12b, 700 feet), and El Sendero Luminoso (V 5.12d, 1,750 feet) in El Potrero Chico, Mexico, on January 2014.* How his experience and all of these various mental, physical, and psychological elements interact would provide clues as to how a brain operates. Answers may come one day after elucidating the network learning paradigms, parallel information processing techniques, underlying functional modularity, and dynamic connectivity in the human brain that are innovations of biological evolution.

A modular, three-layer architecture, such as was developed for the winning vehicle in the 2007 DARPA challenge—Tartan Racing's Boss from Carnegie Mellon University—resonates at least on an operational level with the rock climber's complex, destination-seeking scenario. Figure 8.3 shows a schematic of the neuro-mechanical control system that gets Alex up the wall. The physical hardware comprises the least abstract, bottom layer of the architecture, featuring the climber's limbs and musculature, sensors, and internal GPS. A second layer can be seen as responsible for perception and retains a model of the environment. The perception layer of course resides in the brain with a module that processes sensory data and performs real-time feature detection (positions of rock ledges, distance estimates, for example). A mental model

---

* In North American rock climbing, route difficulty is graded by a three-part code known as the Yosemite Decimal System. Routes are assigned a grade (V routes are typically multi-day roped climbs). Class ranking indicates risk (a class 5 denotes a route where risk of serious injury or death can occur for un-roped climbs). Technical difficulty was historically scaled from 1 to 10, with the single hardest move indicated by the a, b, c, d suffix (d is hardest). On Yosemite, Freerider is rated V 5.12d, which is the route Alex Honnold free soloed in 3 hours and 56 minutes.

of the environment contains the map details and other information that is stored in memory, including a representation of Alex's body in space and time. The perception module interfaces with the third layer, which contains all of the cognitive abilities for planning and execution of movement. Here, the modules consist of a route planner, largely in memory but with the capacity for updating due to obstacles. Next is a "game state" module with situational awareness that tracks what needs to be done next, given the current state. The critical remaining module for the rock climber is his on-the-fly planning and execution of movements, a motion planner. A move, or sequence of moves, gets completed by controlling the physical hardware, and the "monopoly game board" is updated to reflect the new state of play, where along the planned route Alex is on the granite wall.

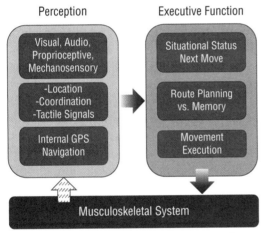

Figure 8.3: From robotics to rock climbing: adopting a modular, three-layer architecture to model motor control

The robotics community found this type of three-layer architecture to be quite successful in staging sequential navigation tasks for robotic vehicles. For AI researchers, the mastery of Atari games came through a different route—the creation of a reinforcement learning AI agent. There were several reasons why Atari games could be conquered with a simpler AI architecture, including the fact that the action space is discrete, there are deterministic rules about play that could be learned in simulations, and the end goals are clearly defined. In more complex scenarios, it is less certain how this sequential model would fare to model behavior. Exploration-based learning of the environment without a roadmap, such as what animals do in the wild, is an altogether different challenge. Producing human-level performance in exploration games, such as *Montezuma's Revenge* and *Private Eye,* have proved difficult for those AI agents from DeepMind.[22]

The architecture underlying the Boss robotic vehicle takes a page from another neuroscience domain—a classical view of the brain's operation from the cognitive sciences. In this framework, the modules correspond to perception, cognition, and motor control, as shown in the upper layers of the model in Figure 8.3. The origins of this framework are from cognitive psychology and a view of the brain as an information processing system, circa 1950. Cognitive neuroscience has worked off of this framework ever since, attempting to build models and understand complex behavior through the mapping of functions onto particular brain structures. Brain regions involved with perceptual processing take external sensory data and create an internal representation that is a reflection of the environment; cortical areas at the next stage perform cognitive functions and produce an understanding of the world and formulate a plan of behavior; motor circuits then execute the desired action.

The simple architecture depicted in Figure 8.3 is conceptually satisfying and powerful for robotic vehicle navigation, but difficult to reconcile with the neuroanatomy of the mammalian brain. Examining the brain's wiring diagram for motor control, as depicted in Figure 8.4, illustrates one of the problems. The mammalian brain was not engineered but evolved over millions of years. Over the course of evolution, simple neuronal circuits graduated to complex interconnected networks, exemplified by the structured layering and hierarchies established throughout the cerebral cortex in mammals. Neural adaptations thus appeared over time, in the context of existing circuitry constraints, to provide functional specialization for new behaviors. An evolutionary perspective of brain and behavior was first echoed by Charles Darwin, who stated: "In the distant future I see open fields for far more important researches. Psychology will be based on a new foundation, that of the necessary acquirement of each mental power and capacity by gradation." Evolutionary psychology is based around the theory that acquisition of cognitive functions occurs gradually via evolution. Modern neuroscience has proponents of such an ethological framework, such as Paul Cisek, who looks to build biologically plausible models of behavior by a process he calls phylogenetic refinement.[23,24] Neuroscience offers plenty of challenges to AI to figure out the computing and networking rules to arrive at a theory of the brain.

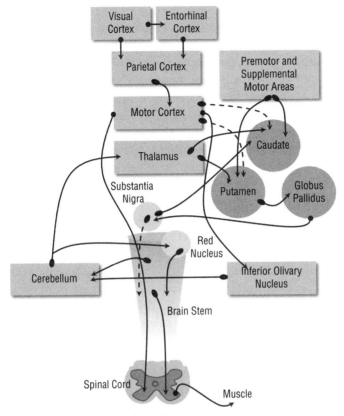

Figure 8.4: Motor control circuitry

The schematic depicts many of the brain areas and pathways known to be involved in motor intention (posterior parietal cortex), modulation (basal ganglia), and movement initiation (primary motor cortex) in the mammalian nervous system. Divisions of the basal ganglia are shown as a cluster of three circles on the right. There are numerous descending and ascending pathways from cortical and subcortical regions, passing through the brain stem and continuing into the spinal cord. Projection of upper motor neurons from the motor cortex via the corticospinal tract terminate at different levels of the spinal cord and connect to neurons in the ventral horn (dark grey region in spinal cord). Lower motor neurons in the ventral horn project out to the limbs to control muscular contraction.

The motor control regions of the human brain offer some of the most stunning examples of the convergence of AI and neuroscience toward the goal of attaining an understanding brain and behavior. The practical applications of basic science research in this area are far-reaching, starting with the recent arrival of brain machine interfaces (BMIs) for paralyzed individuals. The technology already in development is headed toward brain-embedded devices that will read and transmit a person's thoughts to other machines and devices, or perhaps, other people. The use of microelectrode array technology for cortical recording at the single neuron level and AI-based pattern recognition and decoding are revealing much about the representation of motor intentions in the brain. The implanted probes possess 100 to 200 microelectrodes for recording across a one-millimeter squared area of cortex. These measure simultaneous activity of a few hundred neurons with resolution sufficient to resolve spiking activity from individual neurons. Neurophysiological measurements taken from implanted electrodes over the posterior parietal cortex (PPC), the premotor area known as the precentral gyrus, and the primary motor cortex have uncovered the extraordinary ability of single neurons to encode goals, movement trajectories, and movement types.[25–27]

The foundational research on motor control that led to studies in humans was done with nonhuman primates. This prior work laid out a remarkable map—one of intentions—for planned motor actions that showed specificity at the single-neuron level. For example, there are individual neurons in the brain that are selective for the types of movements required for the rock climber, such as the intention to grasp. Intention specificity is also localized in various parietal regions for movements planned for the eye, hands, and limbs. Decoding an individual's intentions is a promising way to build assistive devices, as signals from the PPC directly precede and activate the initiation of motor action by neurons in the motor cortex.

Some of the first results from live recordings in humans were performed on a tetraplegic patient in a study led by Richard Andersen and colleagues at Caltech and USC published in 2015.[25] To determine the brain locations prior to electrode implantation, the researchers used functional magnetic resonance imaging (fMRI) during a period when the individual was asked to imagine reaching and grasping. With these imagined movements, fMRI mapped separate regions in the left hemisphere of the PPC to these activities, where probes were subsequently implanted. Movement intention signals were recorded in the brain during an experiment designed to identify distinct intentional activities during motor imagery. The study captured the neural encoding of the goal of the movement, which was to reach for a target in a spatial quadrant. The multistage task also detected the trajectory encoding. From a series of experiments, the researchers provided evidence for the first time that human neurons in the PPC can encode motor intention signals (goals) or trajectories exclusively or have the ability to encode both. Within this region, neurons also appeared to signal imagined motor activity specific to body regions, such as left arm versus right arm movement. These data allow for machine learning-based decoders to be built into neural prosthetics, where the decoded intent signals can be sent to robotic arms to control movement.

Investigations with microelectrode arrays in regions adjacent to the primary motor cortex have overturned the orthodoxy on how body regions are mapped across motor cortical areas. Krishna Shenoy's team at Stanford University and colleagues from the BrainGate consortium discovered that instead of distinct topographic mappings representing head, face, arm, and leg movements, these regions are mixed in the premotor area.[26] The recording data provided the first evidence of a whole body tuning code, with a region once thought to retain only an arm/hand specificity actually interlinking the entire body. The data led the team to develop a compositional coding hypothesis, which can account for how a neural network might perform both limb and movement encoding. An interesting and testable idea regarding motor skill learning and transfer also emerged from this compositional model. Once a limb had acquired a motor skill, a change in the limb code dimension would accomplish a skill transfer to the nonskilled limb. The team built a neural network model to test the implementation of skill transfer using the compositional representation of movement. The AI-based RNN model validated the code's potential in skill transfer. Components of this compositional code for BMIs are now in clinical trials in BrainGate2 (see `https://www .clinicaltrials.gov/ct2/show/NCT00912041`).

There are now several spectacular examples of real progress being made in the melding of neuroscience and AI. The same group at Stanford has shown that imagined handwriting can be captured by electrodes, decoded by AI algorithms, and used in text composition.[27] Control of robotic arms driven by BMIs continues in the human primary motor cortex in analogous fashion to the studies in premotor areas. Companies such as Neuralink are manufacturing high-density arrays with 1,000 electrodes. Although much of the research has for practical purposes focused on accessible, surface cortical regions, strategies to record or transmit signals to deeper regions are on future roadmaps of researchers and device manufacturers alike.

Engineering Medicines with Biotechnology and AI

What is clear is that evolution is an innovation machine, and nature's products are ready to be let loose to take on new functions, under the discerning eye of the breeder of molecules.[28]

—Frances Arnold, 2018 Chemistry Nobel Laureate

The convergence of the life sciences, engineering, and computer sciences is spawning a raft of technological approaches for creating new medicines. Previous chapters of this book have introduced technologies from these disciplines and their applications either alone or in combination for biological discovery and innovations in diagnostics and therapeutic development. The biotechnology industry is creating a new generation of AI-driven tools, medical devices, and engineering platforms, using new ideas and building on the learnings taken from first- and second-generation technologies. As shown in Figure 8.5, innovation around engineering medicines is now in full swing for molecular design and discovery, genome engineering, gene and cell therapy, digital therapeutics, network engineering, and neurotechnologies. Tools such as machine learning–guided protein engineering follow in the footsteps of generative chemistry, promising to leverage data obtained from laboratory experiments with knowledge-based models to produce proteins with therapeutic potential.[8,28] Molecular design includes protein evolution techniques, which will have wide applications across many industries and in pharmaceutical manufacturing. Beyond new biologics, novel structures are being designed for drug delivery vehicles, nanomaterials, and biocatalysts, and they will also serve as building blocks in synthetic biology workflows to produce microbes and microbial chemical factories.

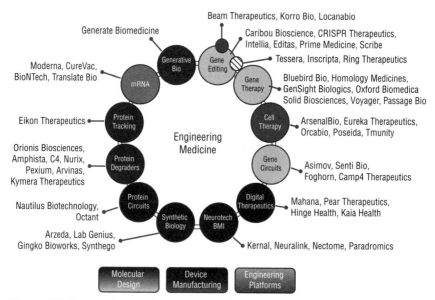

Figure 8.5: Engineering medicines: the new therapeutic development landscape

The trend to develop new therapeutics based on engineering principles is depicted by an array of exemplar companies positioned across three innovation divisions in biotechnology: molecular design, engineering platforms, and device manufacturing. Molecules and devices produced by these new platforms should emerge in five to ten years to impact the industry's overall productivity of therapeutics and complement pharmaceutical pipelines with traditional small molecule-based drug discovery. For gene editing, technologies are being explored for using RNA editing (Beam Therapeutics, Korro Bio, Locanabio) and non-CRISPR genome modification (Tessera, Inscripta, Ring Therapeutics).

The human genome encodes approximately 25,000 proteins, a significant fraction with no known function. Within an average cell, there are an astonishing 10 billion individual proteins at work. Protein interaction networks, protein dynamics, and other properties are poorly charted at this juncture in biology. Rapid proteomic profiling technologies such as mass spectrometry do not provide the full depth of data that is generated with genomic sequencing-based methods. Exploration of the proteome has really only just begun.

Biotechnology's first foray into medicine was to produce therapeutic amounts of well-understood human proteins such as insulin and a variety of antibodies (reviewed in Chapter 4). With the maturation of mRNA platforms from Moderna and BioNTech, a completely new avenue has opened to design, produce, and deliver information for manufacturing vaccines and therapeutics in cells. The most important next stage is the routine design of proteins and computational methods to predict their biochemical function, medical use, or disease-modifying potential. Computational power and AI are now enabling prediction of both protein structure and function from the linear sequence of DNA or string of amino acids. Several groups have implemented architectures based on RNNs for classification of protein function directly from primary sequence. These AI systems are trained on annotated datasets from UniProt that contain more than 20 million protein sequences.[7] The value of high throughput sequencing and omics data is quite clearly compounding over time, as the exponential increase in available data for AI models to train on is fueling a whole new set of capabilities for molecular design.

The ability to extract functional information from raw data and then predict functions of engineered proteins is at the heart of Generate Biomedicine's platform. Advances in microscopy are enabling the elucidation of biological structure down to near-atomic resolution with cryo-EM. The development of super resolution microscopy gives researchers a tool to look more closely at protein dynamics, a platform component of Eikon Therapeutics. A cohort of startups are entering into the race to develop protein degradation platforms. The thesis underlying this approach is that many protein targets are inaccessible or undruggable by small molecule approaches and can be dragged to cellular protein degradation pathways by novel degrader molecules. Some of the companies at the vanguard are Kymera Therapeutics, C4 Therapeutics, and Arvinas, to name a few entrants. Other startup hopefuls in the proteomics space are Octant, which is investigating drugs that modulate multiple targets or pathways, or Nautilus biotechnology, which is developing screening technology to assess G-protein coupled receptor (GPCR) signaling pathways at scale.

The landscape of genome engineering platforms has expanded with a steady proliferation of startups taking new steps to reverse gene mutations, control gene regulation, and correct defective RNAs. Beam Therapeutics is developing therapeutic approaches with a variety of engineered editors, which include the use of DNA and RNA base editors, Cas12b nuclease technology, and prime editors. Others in the RNA editing field are Korro Bio, which is working on adenosine deaminases acting on RNA, known as ADARs. Korro's approach is to use a cell-native editor to get around the requirement for delivery of a CRISPR/Cas system into tissues and cells. Locanabio is also innovating around RNA targeting for gene therapy using RNA binding protein systems. A new genome engineering technology that harnesses the biology and machinery of mobile genetic elements is being pioneered by Tessera Therapeutics. Another novel approach in this arena is being taken by Ring Therapeutics, which has plans to engineer medicines into virus particles that are resident in normal cells and are known collectively as the *commensal virome*. The next bold outsider in biotechnology might well be a computationally minded entrepreneur with a Turing startup to rival Tesla, engineering the medicines of the future.

Notes

1. Schwab K. The fourth industrial revolution. New York: Crown Business; 2017.

2. Pauling L, Corey RB, Branson HR. The structure of proteins: Two hydrogen-bonded helical configurations of the polypeptide chain. PNAS. 1951 Apr 1;37(4):205–11.

3. The AlphaFold Team. AlphaFold: a solution to a 50-year-old grand challenge in biology. 2020. Available from: `https://deepmind.com/blog/article/alphafold-a-solution-to-a-50-year-old-grand-challenge-in-biology`

4. Senior AW, Evans R, Jumper J, Kirkpatrick J, Sifre L, Green T, et al. Improved protein structure prediction using potentials from deep learning. Nature. 2020 Jan;577(7792):706–10.

5. Balakrishnan S, Kamisetty H, Carbonell JG, Lee S-I, Langmead CJ. Learning generative models for protein fold families. Proteins: Structure, Function, and Bioinformatics. 2011;79(4):1061–78.

6. Rives A, Meier J, Sercu T, Goyal S, Lin Z, Liu J, et al. Biological structure and function emerge from scaling unsupervised learning to 250 million protein sequences. PNAS. 2021 Apr 13;118(15).

7. Rao RM, Meier J, Sercu T, Ovchinnikov S, Rives A. Transformer protein language models are unsupervised structure learners. Synthetic Biology; 2020 Dec.

8. Alley EC, Khimulya G, Biswas S, AlQuraishi M, Church GM. Unified rational protein engineering with sequence-based deep representation learning. Nature Methods. 2019 Dec;16(12):1315–22.

9. Fudenberg G, Kelley DR, Pollard KS. Predicting 3D genome folding from DNA sequence with Akita. Nature Methods. 2020 Nov; 17(11):1111–7.

10. Li W, Wong WH, Jiang R. DeepTACT: predicting 3D chromatin contacts via bootstrapping deep learning. Nucleic Acids Res. 2019 Jun 4;47(10):e60.

11. Trieu T, Martinez-Fundichely A, Khurana E. DeepMILO: a deep learning approach to predict the impact of non-coding sequence variants on 3D chromatin structure. Genome Biology. 2020 Mar 26;21(1):79.

12. Schwessinger R, Gosden M, Downes D, Brown RC, Oudelaar AM, Telenius J, et al. DeepC: predicting 3D genome folding using megabase-scale transfer learning. Nature Methods. 2020 Nov; 17(11):1118–24.

13. Quang D, Xie X. DanQ: a hybrid convolutional and recurrent deep neural network for quantifying the function of DNA sequences. Nucleic Acids Research. 2016 Jun 20;44(11):e107–e107.

14. Alipanahi B, Delong A, Weirauch MT, Frey BJ. Predicting the sequence specificities of DNA- and RNA-binding proteins by deep learning. Nat Biotechnol. 2015 Aug;33(8):831–8.

15. Zhou J, Park CY, Theesfeld CL, Wong AK, Yuan Y, Scheckel C, et al. Whole-genome deep learning analysis identifies contribution of noncoding mutations to autism risk. Nat Genet. 2019 Jun; 51(6):973–80.

16. Zhou J, Theesfeld CL, Yao K, Chen KM, Wong AK, Troyanskaya OG. Deep learning sequence-based ab initio prediction of variant effects on expression and disease risk. Nature Genetics. 2018 Aug;50(8):1171–9.

17. Kriegeskorte N. Deep Neural Networks: A New Framework for Modeling Biological Vision and Brain Information Processing. Annu Rev Vis Sci. 2015 Nov 24;1:417–46.

18. Bengio Y, LeCun Y. Scaling Learning Algorithms towards AI. In: Large-Scale Kernel Machines. MIT Press; 2007.

19. Richards BA, Lillicrap TP, Beaudoin P, Bengio Y, Bogacz R, Christensen A, et al. A deep learning framework for neuroscience. Nat Neurosci. 2019 Nov 1;22(11):1761–70.

20. Cobb M. The Idea of the Brain. New York, NY, USA: Basic Books; 2020.

21. Hassabis D, Kumaran D, Summerfield C, Botvinick M. Neuroscience-Inspired Artificial Intelligence. Neuron. 2017 Jul;95(2):245–58.

22. Gerrish S. How Smart Machines Think. Cambridge, MA: MIT Press; 2018.

23. Cisek P. Resynthesizing behavior through phylogenetic refinement. Atten Percept Psychophys. 2019 Oct;81(7):2265–87.

24. Cisek P, Kalaska JF. Neural Mechanisms for Interacting with a World Full of Action Choices. Annual Review of Neuroscience. 2010;33(1):269–98.

25. Aflalo T, Kellis S, Klaes C, Lee B, Shi Y, Pejsa K, et al. Decoding motor imagery from the posterior parietal cortex of a tetraplegic human. Science. 2015 May 22;348(6237):906–10.

26. Willett FR, Deo DR, Avansino DT, Rezaii P, Hochberg LR, Henderson JM, et al. Hand Knob Area of Premotor Cortex Represents the Whole Body in a Compositional Way. Cell. 2020 Apr;181(2):396-409. e26.

27. Willett FR, Avansino DT, Hochberg LR, Henderson JM, Shenoy KV. High-performance brain-to-text communication via handwriting. Nature. 2021 May;593(7858):249–54.

28. Arnold FH. Directed Evolution: Bringing New Chemistry to Life. Angew Chem Int Ed Engl. 2018 Apr 9;57(16):4143–8.

Glossary

A

Active learning　Iterative learning process that makes use of machine learning algorithms to determine the next set of experiments to be carried out while studying a given problem. Uses include determination of the optimal training data from a model, reducing the amount of data required in supervised learning.

Aliphatic　Organic compounds or functional groups that do not contain aromatic rings. Aliphatic hydrocarbons are composed of carbon and hydrogen joined with straight or branched chains, or nonaromatic rings.

Alkaloid　Class of organic compounds with one or more nitrogens in a heterocyclic ring. Caffeine, morphine, and cocaine are alkaloid substances.

Aromatic　Organic compounds that contain a benzene ring.

Artificial general intelligence (AGI)　Machined-based artificial intelligence characterized by human-like intellectual capacities to solve problems broadly, demonstrate creativity, and possess reasoning ability.

AUC   Short for area under the ROC curve. AUC is an important metric in binary classification. The area under the curve is found by plotting the true positives rate (usually y-axis) against the false positive rate (x-axis). This is also referred to as AUROC and as a receiver operating characteristic curve (ROC).

B

Backpropagation algorithm   A supervised neural network learning algorithm that efficiently computes error derivatives with respect to the weights by passing signals through the network in reverse in order to iteratively minimize the error. This is one of the most important and effective algorithms employed in deep learning.

Binary threshold unit   A neuron or unit in an artificial network where the units take on only two different values for their states. The unit value is determined by whether the unit input exceeds its threshold.

Biologics   A class of therapeutics from natural biological sources. Biologics are often composed of proteins and, by definition, are not chemically synthesized, in contrast to small molecule organic compounds. Examples of biologics are monoclonal antibodies, human growth factors, vaccines, gene therapies, and human recombinant enzymes.

Biomarkers   A biomarker is a measurement of biological entity that is used in drug development decision-making. Biomarkers take any one of several types, including molecular, histologic, radiographic, or physiologic characteristic. These can be measured singly or as a composite to provide an interpretive output.

C

Chemical space   A reference to a grouping, or space, of molecules in terms of their physicochemical properties, such as size, shape, charge, and hydrogen-bonding potential. This can refer to enormously large collections of molecules that can exist in theory but may not be available synthetically or obtainable in a physical chemical library.

Classical AI   Refers to symbolic artificial intelligence methods that were under study prior to more recent neural network and deep learning methods.

Clinical trials   A type of clinical research that uses volunteers (in human studies) and designed to study the safety and efficacy of new therapeutic interventions or medical products.

Cohorts   In human genetics, groups of individuals who share a common trait or demographic, such as male diabetics older than 60, that are studied over time.

Computer-assisted diagnosis (CAD)   Medical diagnostic procedure that involves use of computer programs for guiding or augmenting clinical decision-making.

Connectionists   Refers to those cognitive scientists or others in AI who seek to explain intellectual abilities using artificial neural networks modeled after brain connectivity or circuitry.

Convolutional neural networks (CNNs)   A type of deep neural network that contains at least one convolutional layer. These types of networks are widely used in image recognition and have a hierarchical architecture. In a CNN, convolutions are performed on the previous layer with a number of weight-template patterns.

D

Deep learning   A subfield of machine learning that employs artificial neural networks with many layers to identify patterns in data. In deep learning, various machine learning techniques use layers to provide multiple levels of abstraction and learn complex representations of the data (such as CNNs). Learning is accomplished typically by using stochastic gradient descent by error backpropagation.

Digital health   Healthcare enabled by digital technologies across a broad range of categories such as software, mobile health, health information technology, wearable devices, telehealth, and personalized medicine.

E

Epigenomics   A field of biological research to study epigenetic changes such as DNA methylation, histone modification, chromatin structure, and noncoding RNA using molecular profiling techniques.

Expert systems   A class of computer programs built with artificial intelligence methods to emulate human expert behavior and knowledge for completion of a task.

F

Forward synthesis   The sequential steps in organic chemistry undertaken to build compounds from initial chemical building blocks.

H

High-throughput screening (HTS)   A general term in biological lab or drug discovery settings that refers to a systematic approach to evaluate chemical compounds, genes, or biomarkers in a simple assay format amenable to high throughput.

Hit discovery   The early phase in drug discovery to identify small molecules that have some desired activity in a compound screen against a target.

Horizontal gene transfer   The process in bacteria where genes or genetic information are acquired from contemporary organisms in the environment rather than from cell division.

I

Immunogen   An immunogen is a substance that induces an immune response when entering the body. This is also referred to as an *antigen*.

In silico* biology   The study of biological processes and phenomena by simulation and modeling with computational methods.

K

Knowledge-based systems   A type of artificial intelligence where computer programs use a knowledge base to construct or build solutions to problems. The input knowledge aids in reasoning for the system.

L

Ligand-based virtual screening   In drug discovery, a screening that employs computational tools and information around the ligand instead of the target to assess and identify promising lead chemical compounds. This approach is an alternative when target 3D structures are absent or uninformative for screening programs.

M

Medicinal chemistry   Branch of chemistry concerned with design and characterization of compounds for pharmaceutical development.

Middle Eastern Respiratory Syndrome (MERS)   A viral respiratory disease caused by MERS-CoV, a coronavirus discovered in the Middle East in 2012.

Molecular targets   The focus of a drug discovery program, usually chosen based on experimental results or genetics data and that can be targeted by small molecule organic compounds to interfere with a disease process.

Monoclonal antibody technology   Molecular and cellular techniques for generating and producing antibodies for a specific target site on an antigen.

Multitask learning   A strategy in machine learning for helping a model improve by using knowledge contained in multiple, related tasks. This is used in a variety of settings, such as multitask active learning, multitask supervised learning, and multitask reinforcement learning.

N

Natural language processing (NLP)   The analysis, processing, and interpretation of natural language by computer programs and other systems to extract meaning or content. The NLP domain spans multiple disciplines, including linguistics and computer science, and it uses language processing technologies such as speech recognition, text classification, machine translation, and dialog systems.

Neural networks   In computing, a model that is composed of layers consisting of simple connected units or neurons followed by nonlinearities. Neural networks have at least one hidden layer.

New molecular entity   A medication containing an active ingredient that has not been previously approved for marketing in any form in the United States. NME is conventionally used to refer only to small-molecule drugs.

Nucleosides   Organic compounds formed from a nucleoside base and a five-carbon sugar. Nucleosides do not contain a phosphate group, as found in nucleotides.

P

Phage display   A molecular biology technique for screening and capture of ligands. The technology uses bacteriophage components in cell-based systems that can express peptide or protein sequences on the cell surface as fusions of phage coat proteins.

Pharmacodynamics   A term in pharmacology that refers to the effect of a drug on human physiology.

Pharmacokinetic   A term in pharmacology that refers to the body's effect on a drug via metabolic processing and other actions.

Pharmacopeia   Reference texts published for collections of medicinal compounds and containing recipes or directions for preparation. Also this is used to refer to a set collection, either in nature or from a special grouping or study.

Picture Archive and Communications Systems (PACS)   The digital documentation systems for medical images found in hospitals or medical centers.

Point mutation   A change at a single nucleotide position in DNA or RNA sequences. There are two classes of point mutations. Transversions interchange a purine for a pyrimidine base, such as when a single purine (A or G) is changed to a pyrimidine (T or C), or vice versa. Transitions occur between two purines (for example, A to G) or between two pyrimidines (for example, C to T).

Q

Quantitative structure-activity relationship (QSAR)   In drug discovery and design, an approach to define the relationship between the structural features and physiochemical properties of a compound and its biological activity.

R

Recombinant DNA technology   Molecular biology techniques to construct and propagate new DNA molecules, typically by introducing chimeric constructs into bacteria using plasmids as vectors. They are generically referred to as gene cloning but encompass a broader range of goals to move or manipulate DNA or express genes in organisms.

Reinforcement learning   A machine learning paradigm adopted from neuroscience that employs algorithms that learn with an optimal policy. The policy operates on a goal—to maximize return when interacting with an environment, such as a game. Reinforcement learning systems evaluate sequences of previous steps that led to gain or loss toward the goal in order to learn.

Retrosynthesis   The synthesis of complex organic compounds by a process that works "backward" from the desired product. Synthesis steps are proposed in sequence that will provide the building blocks for the compound.

RNA interference (RNAi) A gene silencing mechanism that acts post-transcriptionally on mRNAs to target recognized molecules for destruction. Short, double-stranded RNAs produced by the cell recognize homologous sequences as part of the RNAi mechanism.

S

Semi-supervised learning A machine learning technique where a model is trained on data with an incomplete set of labels. The technique exploits the fact labels can often be inferred for the unlabeled examples and then used to train and create a new model.

Single nucleotide variant (SNV) In genome sequencing, a position in an individual's genome that differs from the organism's reference sequence. Not all SNVs are polymorphic variants, which are known as SNPs and have a certain frequency in a population.

Structure-based virtual screening (SBVS) The use of protein 3D structure to initiate a search for high-affinity ligands in chemical libraries using computational methods in drug discovery.

Supervised learning A main style of machine learning that requires input patterns (categories or labeled data) and additional information about the desired representation or the outputs.

Symbolic AI The use of symbols and logic in systems for machine intelligence.

Symbolists Proponents of symbolic AI that focus on use of rules and symbolic logic in artificial intelligence systems.

T

Toll-like receptors (TLRs) An important class of receptors that detect molecules from foreign organisms. TLRs are type I transmembrane receptors that recognize specific molecular patterns present in substances such as foreign nucleic acid sequences.

Transfer learning A method in artificial intelligence that takes previously learned data and applies it to a new set of tasks.

Turing test A test of artificial intelligence suggested by Alan Turing. The test evaluates a scenario in which a machine's responses in a conversation or interchange with an observer would be distinguishable from that of a human's response.

U

Universal Turing machine   An abstract concept that one machine or device can capture all of "the possible processes which can be carried out in computing a number" (Alan Turing 1936–1937).

Unsupervised learning   A type of machine learning that trains a model with unlabeled data to infer underlying patterns in the dataset.

X

X-Omics   A term for any of the systematic biological research methods used for characterizing the genome (genomics), transcriptome (transcriptomics), proteome (proteomics), metabolome (metabolomics), and lipidome (lipidomics).

Index